Gamechanger AI

Klaus Henning

Gamechanger AI

How Artificial Intelligence is Transforming our World

 Springer

Klaus Henning
Aachen, Germany

The translation was done with the help of artificial intelligence (machine translation by the service DeepL.com). A subsequent human revision was done primarily in terms of content.
ISBN 978-3-030-52896-6   ISBN 978-3-030-52897-3 (eBook)
https://doi.org/10.1007/978-3-030-52897-3

© The Editor(s) (if applicable) and The Author(s), under exclusive license to Springer Nature Switzerland AG 2021
This work is subject to copyright. All rights are reserved by the Publisher, whether the whole or part of the material is concerned, specifically the rights of translation, reprinting, reuse of illustrations, recitation, broadcasting, reproduction on microfilms or in any other physical way, and transmission or information storage and retrieval, electronic adaptation, computer software, or by similar or dissimilar methodology now known or hereafter developed.
The use of general descriptive names, registered names, trademarks, service marks, etc. in this publication does not imply, even in the absence of a specific statement, that such names are exempt from the relevant protective laws and regulations and therefore free for general use.
The publisher, the authors, and the editors are safe to assume that the advice and information in this book are believed to be true and accurate at the date of publication. Neither the publisher nor the authors or the editors give a warranty, expressed or implied, with respect to the material contained herein or for any errors or omissions that may have been made. The publisher remains neutral with regard to jurisdictional claims in published maps and institutional affiliations.

This Springer imprint is published by the registered company Springer Nature Switzerland AG
The registered company address is: Gewerbestrasse 11, 6330 Cham, Switzerland

Preface

During my studies 50 years ago, I had already learned something about *neural networks*. At that time, it was a very exciting discovery for me to be able to reproduce the basic functions of a nerve cell of a living being using a computer program. Such a neural network is shown in simplified form in Fig. 1. It contains many parallel inputs, all of which act on a first hidden layer. This layer consists of nodes and each node receives information from all the available inputs.

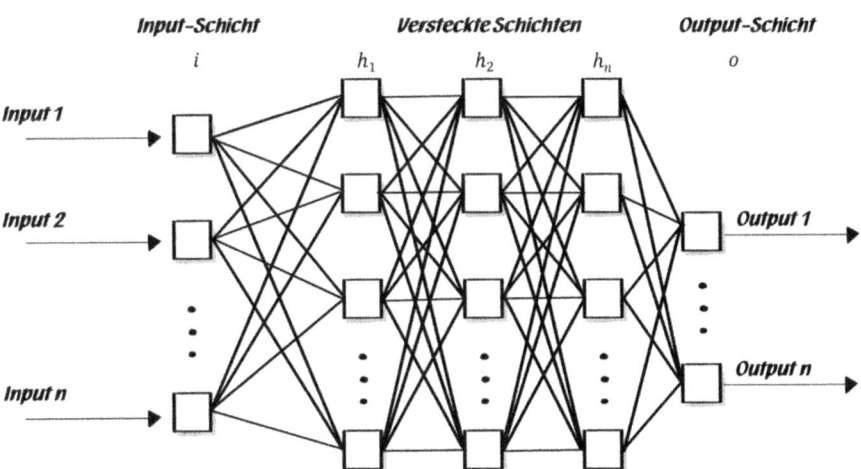

Fig. 1 Representation of the structure of a neural network (https://en.wikipedia.org/wiki/Artificial_neural_network, accessed in April 2020)

Each node processes and weighs this information, passing it on to every node in the following layer. In the end, you end up at an output layer.

This output layer is now ready to be used as an additional input layer. Through these feedback loops, the neuronal network learns from its own results.

These networks of nodes were mirrored by the structures of the nerve cells I had learned about as a student. Each individual node is structured like a nerve cell (Fig. 2).

I remember being impressed by the incredible diversity and ability of nature to deal with information. Every piece of external information is processed with different weightings in each nerve cell and leads to a message, the so-called activation function, which is then forwarded to all nodes of the next layer.

50 years ago, it was clear to us that this was a pretty clever construction with a lot of potential for application. Some renowned scientists predicted a great future for advances following this theoretical framework. However, this proved to be wrong in the following decades. The time had not yet come. It was far too complex technically and therefore seemed unlikely to have any significant effect on ongoing technical development, at least for the foreseeable future.

The reality turned out to be different.

Back then, it gained respect for nature's enormous achievement and its wasteful effort. I learned that the hamstring reflex of the frog's leg alone

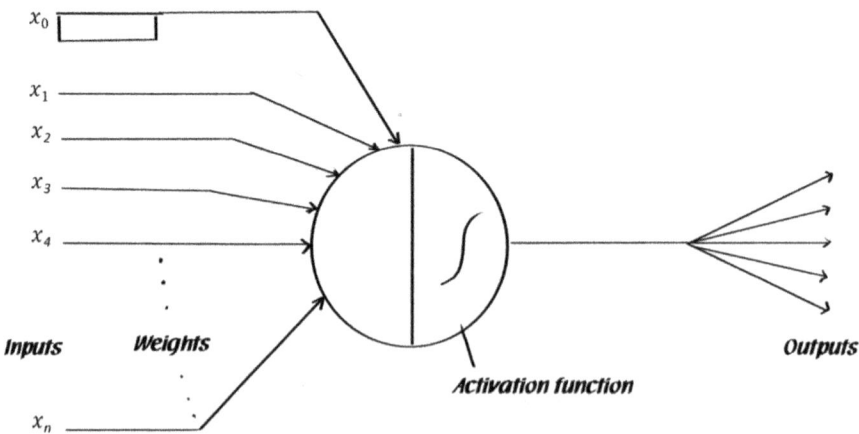

Fig. 2 Structure of the replica of a nerve cell (https://en.wikipedia.org/wiki/Artificial_neural_network, accessed in April 2020)

Preface vii

contains over a dozen highly complex parallel control loops, each full of neuronal pathways consisting of countless layers of neuronal networks.

I came to the following conclusion: With technology, we can make things easier. You don't have to make it so complicated to just stabilize the hamstring reflex in a frog's leg.

Again, reality turned out to be different.

Of course, in my cybernetics lectures 40 years ago, I often told my students about all manner of possibilities. For example, I recently found this sketch in my records, detailing a possibility for the automatic settlement of traffic fines (Fig. 3).

I was of the opinion at the time that this would come very quickly. And although it has been technically feasible for 40 years, to my knowledge there is still no system in which direct deposits of fines are coupled with the devices in one's own home. This would allow all family members to have complete transparency and to get notified immediately which family member drove too fast where.

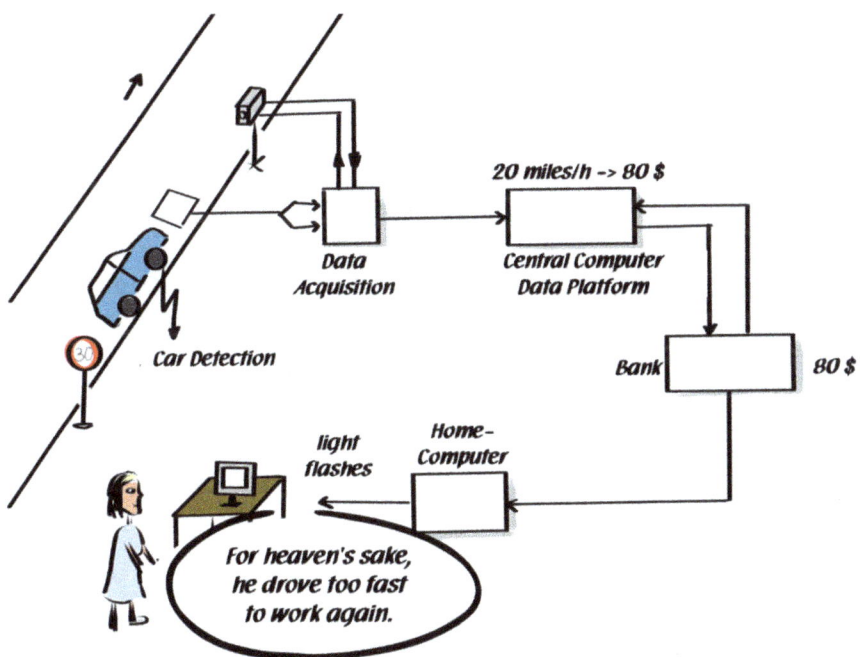

Fig. 3 The automatic fine machine, a vision from 1985 [Henning, Klaus: Kybernetische Verfahren der Ingenieurwissenschaften (Cybernetic Procedures in Engineering Sciences). Mainz, Aachen 1986]

But enough about the past. Artificial intelligence today is a powerful tool whose basic mathematical construction has existed for two generations. Only now does it lead to dramatic changes in the reality of our lives and work.

The continuing increase in computer capacity has made it possible to exchange and process almost unlimited amounts of data worldwide. At the same time, the computers are getting smaller and smaller and here too the end is not yet in sight.

On the other hand, there is always the phenomenon that technical developments are possible but do not prevail and spread. In this respect, any prediction as to when which form of artificial intelligence will penetrate which part of our lives has an enormous degree of uncertainty.

Here is what we can say based on the development so far: When artificial intelligence systems spread, they do so extremely quickly and worldwide. We can observe that in several areas. But when artificial intelligence has to do with the real things of this world, it often takes much longer than expected.

With these preliminary remarks, I now invite you to begin a journey with me. In addition to theoretical facts, I have also written down my personal experiences and assessments in this non-fiction book. The positive attitude towards digital transformation with artificial intelligence corresponds to my conviction. A great deal is written and discussed about the negative and risky aspects. That is why we will not deepen these aspects but focus on the opportunities. On this journey, we will see how this world has been and will be changed by the gamechanger artificial intelligence. The challenge of the coming decades will be to ensure that this change succeeds for the good of mankind.

Aachen, Germany Klaus Henning
May 2020

Acknowledgement

I would like to thank everyone who accompanied me on my journey of this book. Sabina Jeschke helped me to open up to the far-reaching perspectives of artificial intelligence. Stephanie Bauduin has worked tirelessly on details of content and form and has designed most of the pictures. Many others have contributed to the success of the book through their intensive feedback—Andrea Heide, Tobias Meisen, Robert Henning, Max Haberstroh, Teresa Merz, Thomas Bergedieck, Susann Morgenstern, Rainer Bernhardt, and Renate Henning. Many thanks to Laura Bergedieck, Rodrigo Guzman-Sanchez in Chicago, and Shari Holte from umlaut transformation for revising the English translation.

Thanks also to the AI deepL.com translating machine. It provided an excellent basic translation of the English version of this book in less than 2 min.

Contents

1. It's All About Us 1
2. The Objects of This World Become Intelligent 9
3. How Did Artificial Intelligence Come into Being and Where Do We Stand Today? 21
4. Can Machines Have Their Own Consciousness? 33
5. The Homo Zappiens 43
6. The Inverse Gutenberg Revolution 53
7. The Age of Hybrid Intelligence 61
8. The Digital System Landscape 71
9. On the Way to New Business Models 79
10. Artificial Intelligence Is a Gamechanger of All Jobs 85

11 Everything Is Linked to Everything
    and Becomes Transparent. 91

12 The Ethical and Legal Implications 99

13 Guidelines for the Necessary Redesign of Our Regulatory
    Systems in Industry and Society 107

14 Epilogue: Does Artificial Intelligence
    Make God Redundant?. 119

About the Author 123

Bibliography. 125

Index. 129

1

It's All About Us

Contents
AI Is a Gamechanger.. 2
The Age of Digital Transformation Has Begun... 6

Welcome to an excursion into the age of digital transformation and the gamechanger artificial intelligence (AI).

I sit in the middle of a snowstorm in a cozy chalet in the Swiss mountains and look banned at my birdhouse, which I have just filled with fresh food. The thermometer is showing 10 degrees Fahrenheit. However, it doesn't take 10 min for the first bird to fly in and discover the food. It then takes another 10 min until about 20 more birds fly around the feeding all at once.

Suddenly, it shoots through my head: How would 20 small drones, equipped with systems of strong artificial intelligence,[1] try to get the food from the feeder without human intervention? I observe the speed and agility of the birds, their seemingly chaotic strategy of approaching the feeder, and come to the following conclusion:

By the time we have reached the point where all 20 drone systems will empty the feeder, collision-free, in a very small space, without external input, and with the same speed and agility as these birds, it will take quite a while, certainly more than a generation.

As long as these kinds of interactions "only" take place in virtual spaces (like the internet), making them work is still relatively simple. But when it

---
[1] https://en.wikipedia.org/wiki/Strong_AI, accessed in April 2020.

comes to introducing artificial intelligence systems into "physical reality," it becomes arduous. The "last mile" to realize complex AI support for mechanical systems is especially difficult and tedious.

Until AI systems in such devices as drones have the intelligence, agility, speed, and dexterity of these birds, a lot of work in research and development is still required.

But the world is working on it. For example, a major aircraft manufacturer is in the process of designing a parcel center equipped to handle 10,000 shipments per day. All shipments are to be carried by drones. This would require about five take-offs and landings per minute. The coordination problem is gigantic from a software design perspective—not even considering that the "small side problem" of loading drones with fully automatic AI-controlled transport robots in such a confined space is not yet solved.

And then there is the problem of air traffic control when so many drones are buzzing through the airspace. Already, a Silicon Valley group of companies is trying to simulate how such an AI-controlled system might work.

My message to the reader is twofold:

> **If one goes into detail, the implementation of artificial intelligence systems (AI systems) is extremely difficult and laborious. If it succeeds, however, there will be radical breakthroughs across worldwide applications in an extremely short time.**

Rapid dissemination will be accelerated, when usefulness is proven, and people waive all privacy concerns because of the benefits they receive. You might raise an eyebrow at this but surely, you usually check off those ubiquitous Terms of Use disclaimers very quickly without reading them, don't you?

AI Is a Gamechanger

When technologies "surprisingly" find a mass application in a very short time and processes, habits, learning processes, and order systems are turned upside down, we speak of a so-called disruptive innovation—a gamechanger.

It is often assumed that disruptive innovations are a new phenomenon only when they occur in connection with the internet, digital transformation, and artificial intelligence.

Yes, the digital transformation of our lives is a dramatic transformation. But is such a revolution so unique in the history of mankind?

Let's travel back in time together.

Around 1750, the first industrial revolution began with the invention of the steam engine, i.e. the systematic use of water and steam power. This became the basis of mechanical production.

The next milestone came at the beginning of the nineteenth century through electrical energy. This allowed energy to be transported to any location. This structure laid the foundation for mass production and division of labor. At the beginning, in a period of only 10 years from 1903 to 1913, this enabled the change from horse-drawn carriage mass transport to car mass transport. In the beginning, the first drivers were fined because they had exceeded the speed limit intended for horse-drawn carriages.

Another 70 years later, the digital revolution began, initially "only" related to computer technology and communication technology.

Only today do we feel the full extent of this digital revolution, because it covers the information revolution, in which everything is connected with everything and the world begins to grow together into one huge "brain." Autonomous systems and systems of autonomous systems are growing across a worldwide network, in both factories and administrations. And here's the bottom line: What seemed 50 years ago as a strange dream of computer science can suddenly be realized[2].

> Machines and systems have—at least at a low level—their own consciousness and can independently determine for goals and solutions that nobody has taught them before.

The special thing about it is that the objects of this world are linked with each other as in the "Internet of Things" (IoT), all around the globe.

Today, there are already vehicles with the so-called autopilots, in which the experiences of the vehicles with road courses and curves are exchanged overnight among all vehicles of this class all over the world. They are already learning valuable lessons from each other through a worldwide network. With the enhanced interconnectivity provided by 5G technology and the power of future quantum computers, these exchanges will be possible in a few minutes.[3]

---

[2] For more information on the concept of consciousness see Chap. 4.
[3] Sabina Jeschke: 3 Shadoes of AI – 5G and Quantum Computing setting the stage for next generation AI. Keynote Wirtschaftspolitik aus erster Hand at Leibniz-Zentrum für Europäische Wirtschaftsforschung (ZEW) (Feb. fifth, 2020), Mannheim. http://www.sabina-jeschke.de/Talks/2020/2020-02-05_ZEW_talk.pdf, accessed in April 2020.

If such an interconnected vehicle drives on a dirt road, it might have trouble navigating through unknown curves. But if you take the same turn a week later, the handling is already quite good, because the car has exchanged its experiences with all other cars of this class in the world. Driving experiences with similar curves were compared with each other. Every vehicle in the world is then better prepared to handle similar curves.

There is no doubt that this upheaval is gigantic. It is in its dimension, in its extent, and in its nature not comparable with the first industrial revolution around 1750 and the energy revolution around 1900.

A lot of people haven't realized that "big bang" yet.

Going back further in our history, you will come across the upheaval the world has undergone since the introduction of mass printing. One might naturally think of Luther and the effects of the reformation. But the roots of this movement began 100 years earlier with a disruptive innovation of Gutenberg's printing press, the effects of which are perhaps most comparable with the dimensions of the digital transformation with artificial intelligence of today (Eisenstein, 2009; Mai, 2016).

> **Over a period of 10 years from 1450 to 1460, Gutenberg turned the world upside down with the invention of the letterpress machines.**

Gutenberg was inspired by an idea. It was the idea of a world being designed by reason and logic. To get this idea off the ground, he was convinced that the power of images had to be abolished and that the power of the written word should be accessible to everyone.

And this required the ability to print a text in large quantities and in the same quality without errors. It only took a decade for the first printing press to become operational. And it was a development that rose almost out of nowhere.

Copper stamping with presses was invented in Nuremberg for military purposes. This made it the basis for the so-called patrize,[4] i.e. the first letter stamp.

Gutenberg was a radical entrepreneur. He needed a lot of money for this development and gradually committed most of his retirement savings to the project.

---

[4] A patrix is the embossing stick during printing, i.e. the counterpart to the printed letter.

The first patrize was a " holy mirror," an ornate "single letter," which he sold in large numbers on the occasion of the pilgrimage to Aachen Cathedral.[5] The rush to touch the holy relics had become so large that the chapter of the cathedral of Aachen decided to sell mirror images of the relics as souvenirs instead. This provided Gutenberg with much needed interim financing.

But his goal was to have movable letters. He invented the type case. He invented the "floating" letter. He invented the letterpress machine. He was the first one to use paper print instead of parchment. He solved the problem of stable and replicable mass production of texts.

Additionally, he had to entice employees away from transcribing texts by hand; an industry that almost no longer existed in Europe a generation after the first mass printing of books.[6]

Gutenberg was a typical production engineer, in love with scaling, to be able to produce everything in high quality in large quantities. He invested and invested— almost regardless any losses.

For his first book, he had chosen a work that has remained one of the world's bestsellers to this day—the Bible. His aim was to publish the book with as few pictures as possible.

And all this at a time when Europe was just about to disintegrate; there were three differing popes simultaneously vying for supremacy within the Roman Catholic Church.

Ultimately, it took him only 10 years, until 1460, to produce this dramatic innovation; and barely 10 years later Europe would be littered with letterpress machines with movable type. Within just 20 years, written text was widely available to anyone, anywhere in Europe. It is estimated that before 1450 barely 10% of monks in the monasteries could read; and scant few could do so outside the monasteries. That's why everything was transmitted through pictures—hence the importance of representations of biblical stories in churches.

Within less than a generation a general literacy was established among European societies. The monopoly of the few aloud to others was broken.

Ironically, Gutenberg became the victim of his own invention when his hometown Mainz was defeated in a local war in which for the first time, leaflets were used on a massive scale to mount an effective campaign of psychological warfare.

---

[5] The Aachen pilgrimage. https://heiligtumsfahrt-aachen.de/en/aachen-pilgrimage/, accessed in April 2020.

[6] In a similar way today translation with AI substitute the major part of the translation business. The basic English version was translated in less than 2 min. https://www.deepl.com/translator, accessed in April 2020.

With the printing press, the technological driver for the transition into the age of reason was born. But first, Europe would undergo a period of serious unrest. One of the most important of which would occur nearly 50 years after Gutenberg's death, when in 1517 Martin Luther published his 95 theses against the selling of indulgences in Wittenberg.

Certainly, Luther was not aware of his long-term impact. There were unwanted feedback processes, remote actions, and side effects. The beginning of the enlightenment was on the rise: the written word and reason were in the ascendancy. Pictures disappeared on a large scale from daily life. Luther may not have intended to promote the split of the church; however, his actions led to secessions from the Catholic Church. Thus, many Christian denominations arose, which were shaped by the primacy of the written word before pictures and myths.

Europe was still not done with its share of strife. As a result of the mental and spiritual unrest, the political order disintegrated, leading to the Thirty Years' War (1618–1648). It would take about 100 years after the breakthrough of reason, made possible by mass printing, for Europe to find its way back to new systems of social and political order.

Back to today:

The Age of Digital Transformation Has Begun

> I am convinced that the biggest disruptive innovation since Gutenberg is taking place today. That's the real big bang many people haven't heard yet.

Like then, the whole social order, the way people live, work, and what they believe is being turned upside down. Back then, it was limited largely to the states of Europe—today, it simultaneously affects all societies, all the way to the farthest corners of the earth (Henning, 2018).

Today's global changes do not only affect digital networking. It's about more than that![7]

---

[7] Henning, Klaus: How artificial intelligence changes our world. http://www.futur2.org/article/wie-kuenstliche-intelligenz-unsere-welt-veraendert/, accessed in April 2020.

- The machines, cars, and objects of everyday life will be given their own consciousness, which, together with that takes place within minutes, represents a completely new dimension.
- This digital transformation into digital agents, digital shadows, digital twins, etc. is a global and local revolution that will radically transform all areas of our lives. It is inevitable.
- The digital universe with intelligent machines and networks poses a huge opportunity to reinvent our living, working, and learning environments.

Today it is artificial intelligence that penetrates all machines, systems, and devices, all offices, and all areas of private life. Machines, devices, digital platforms, and smartphones increasingly have their own consciousness and make their own creative decisions, which nobody has taught them before.

> We now have the chance to shape the digital transformation with artificial intelligence in a responsible way—before others do it irresponsibly. We are in this together; and we still have the time!

2

The Objects of This World Become Intelligent

Contents

Digital Companions	10
Omnipresent and Unobtrusive	11
Useful and Indispensable	11
And What About Data Privacy?	13
A Paperback Encyclopedia Is Outdated Technology	14
Working in the Digital Age	15
Artificial Intelligence Is a Fundamental Gamechanger	17

Since a while now, statistically speaking, every citizen on earth owns a mobile phone.[1,2] But still, only 67% of the world's population has a mobile connection—half of them by smartphone, and 59% have internet access (2020).[3] Nevertheless, it will take another 5–10 years before the smartphone has reached every corner of the earth. Then, almost everyone will have access to everything around the globe—be it for data, international markets, or educational needs.

---

[1] https://en.wikipedia.org/wiki/List_of_countries_by_number_of_mobile_phones_in_use, accessed in April 2020.

[2] Digital 2020. Figures on the global use of smartphones, Internet and social media: https://www.mcschindler.com/digital-2020-zahlen-zur-globalen-nutzung-von-smartphones-internet-und-social-media/, accessed in April 2020.

[3] Digital 2020 Global Digital Overview (January 2020). https://www.slideshare.net/DataReportal/digital-2020-global-digital-overview-january-2020-v01-226017535, accessed in April 2020.

Digital Companions

Over the next few generations, it seems that the pyramid of needs will change. In addition to water, food, and education comes the need for a mobile phone, Wi-Fi, and a charger for electricity.[4]

I recently experienced a total blackout of my smartphone on a trip abroad to the middle of Eastern Europe. That was a real shock: flight tickets, train tickets, my next appointments, telephone, communication of any kind, none of that was possible anymore. I was suddenly completely on my own. That was when I realized just how connected the whole world has become.

That's how it is with our digital companions: They are extremely useful and that's why we use them. Right now, our digital companions are still rather stupid. Once they become intelligent, they will think for themselves and make independent decisions that no one has suggested to them before. They'll do what is allowed, and what is not.

Here is an example: One of our digital companions will be the autonomous car, an intelligent robot on wheels. Such an autonomous car, which thinks for itself, will of course drive faster than 30 in the 30 miles speed limit zone, if everyone else also drives faster. It will follow rules as much or as little as we do. Therefore, we will have to have a traffic offender index for fully automatic cars. Consequently, these cars will also download apps for speed camera warnings. The ethical and regulatory issues involved will lead to many more discussions.

Such a digital companion can also be 3D glasses which instruct the craftsman on how best to mount a window. These glasses present a digital overlay of what I am doing right now, whether I am doing it well, and a simulation of what I have to do next. The system will also help me learn from my mistakes. Besides, it will help me with my weaknesses. If it is well done, a real win–win situation!

> **In the future, these digital companions will be omnipresent, unobtrusively penetrating all aspects of our personal and professional lives.**

Now you may object: With all the risks attached, we must forbid these digital companions to think for themselves. Well, do you really think that it would

---

[4] Dietrich Identity GmbH: The Maslow Pyramid of Needs in Times of Generation Z. https://www.dietrichid.com/wissensartikel/die-maslowsche-beduerfnispyramide-zeiten-der-generation/, accessed in April 2020.

be of any interest to the rest of the world if we prevent machines and equipment from thinking for themselves within the borders of our own country?

Omnipresent and Unobtrusive

Such intelligent objects will become at least as intelligent as some of the more highly developed animals, such as ravens, horses, or dogs. Whether we like it or not.

Some people may now think of science fiction movies. But you don't even have to go that far. Digital real or virtual digital companions will spread step by step, omnipresent and unobtrusive, in many areas. And in certain areas they will simply be capable of doing more than what we humans are capable of. This is because they can collect and process vast amounts of data in the shortest possible time.

Today, many of the newer smartphones already have an internal AI machine to automatically create photo sequences accompanied by appropriate music. This "For You" function has now reached a point where, for example, it reminds you of previous mountain tours with a photo sequence the moment you find yourself in the same area again. The AI system was not ordered separately—it was simply included. Most users of the corresponding smartphones are not even aware that this function exists. The AI system is purely local and not connected with a database in a "cloud"—a virtual mass storage device.

Most people who use the AI "For You" are enthusiastic about it and use it constantly because it is more attractive and entertaining than clicking through many photos manually. And if you don't like the style, you can choose another one, for example, "cheerful" or "epic." However, some of those users don't even realize that there is an AI algorithm working in the background.

AI applications unobtrusively sneak into our everyday lives and will soon be omnipresent. By 2030, much of it will have become "normal" and we will use such intelligent functions in many places.

Useful and Indispensable

These digital companions will have creative ideas and develop new concepts; in short, they will behave in a way they have not learned from anyone. Combining of a lot of data, using neural networks and the increasingly sophisticated algorithms of artificial intelligence, these digital companions will be able to learn independently, reflect on what has been learned, and develop new behavioral strategies without much trouble.

> **Many intelligent digital companions will make their own decisions.**

That is why, in the longer term, objects of this world that work with such strong artificial intelligence[5] must also become legal entities in their own right, as the European Parliament has rightly brought into the debate.[6] There must be ethical standards and values for the algorithms of such digital companions.[7] Then it will be a good thing.

> **We don't need digital fools.**
> **We need digital companions that are capable of being useful partners.**

For example, if you have to commute into the city at 7 am in the morning, it will be no problem at all for an intelligent navigation device to look at traffic and weather data of the last 10 years and then perhaps give you the following recommendation: Wait 5 min before you leave today. Then it will be less stressful, and you won't lose any time. Because the system knows that 5 min later the traffic jam will have decreased to such an extent that you will reach your destination faster. Such an assessment is impossible for us humans. I am sure we will use such technology when it becomes available. The reason for that is:

> **Usefulness triumphs over data protection.**

---

[5] https://en.wikipedia.org/wiki/Strong_AI, accessed in April 2020.
[6] Janosch Delcker: Europe divided over robot "personhood" (Nov. fourth, 2018). https://www.politico.eu/article/europe-divided-over-robot-ai-artificial-intelligence-personhood/, accessed in April 2020.
[7] Committee on Legal Affairs of the European Parliament, 2015/2103(INL) (Jan. 27th, 2017): Recommendation to the Civil Law Rules on Robotics, Liability, AC to AF. https://www.politico.eu/article/europe-divided-over-robot-ai-artificial-intelligence-personhood/, accessed in April 2020.

As such, the useful idiots in our cars and in our pockets gradually become indispensable partners with whom we wouldn't want to live without. Anyone who has ever learned to appreciate the quality of Google Maps' congestion forecasts don't care anymore that users are tracked digitally—who cares where I go? The main thing is to get past the traffic jam ahead of me. The excellent navigational systems of many vehicle manufacturers use exactly this Google data as a foundation. This data then forms a complete picture of user behavior. This is called a digital shadow.

We happily give away our data for the benefit of using such applications. It is of no use believing that I can keep my data to myself anyway. Far too often, I have agreed to give out information about myself free of charge. In most cases, we are tired of reading the general terms and conditions. Have you completely read how your data is handled during your most recent visits to websites on the internet? Or did you just scroll down and checked the box? However, this development has a considerable downside. Those who have the data are gaining more and more control as well as social, economic, and political influence.

And What About Data Privacy?

It will be interesting to see how long we are willing to actually give away our data for free. It is possible that data markets will develop in which I pay for the right to receive a service in exchange for my data, i.e. I enter into a swap. It is undoubtedly problematic if free apps collect data that they do not need for their functionality, only to then resell it to third parties.

In any case, my smartphone, even without any active function, already allows someone to track me anywhere I go.

I noticed recently that I have three navigation devices in operation when driving my car. First the manufacturer's onboard navigation system, then a traffic forecast app on my mobile phone which is also tied into a database of live traffic obstructions and controls. And third, my car is also connected to the manufacturer's service hotline so that I can get help quickly in case of an emergency. Finally, there is also the engine control unit and other control devices in the vehicle that can be used to subsequently read out trip histories.

A worldwide network has already been formed of all the digital shadows of people, machines, vehicles, and devices that communicate very intensively with each other all the time. We put a high value on the comfort of having the train booking system readily available on an app and the ability to get that information in real-time.

> A kind of digital shadow economy has emerged.
> Mass data is the essential basis of artificial intelligence.

No data protection law will be able to stop this development. In case you don't believe that, just listen to what constitutional rights experts have to say. They will tell you which data can be obtained without violating data protection regulations, which are particularly strict in Germany. In an information paper of the Bavarian Office for the Protection of the Constitution it says[8]:

> People still play a large, if not the most important, role in cases of industrial espionage… Most information is unintentionally revealed when dealing with other people.

In this respect, the challenge of data protection is nothing new. There has always been not only industrial espionage, but also, for example, the curious neighbor who is always extremely interested in who does what with whom and when in the village. What is new is the extent to which data can be collected. In that context, we often lack the understanding and appropriate strategies to protect our own data.

A Paperback Encyclopedia Is Outdated Technology

Another example: Which one of us still uses a paperback encyclopedia? That's yesterday's technology. Maybe the older ones among us still use them. But my grandchildren stand amazed in front of an encyclopedia and ask, "Grandpa, what is this? Did you really use something like that"?

---

[8] Elsasser, Thomas: Dangers of industrial espionage. Annual Conference of Xenium AG. Munich 16.10.2015.

Brockhaus, a German publisher,[9] missed this development and was forced to discontinue its encyclopedia series in 2013. Encyclopedias grew out of fashion precisely the moment when confidence in Wikipedia's data became greater than confidence in the data of a paperback encyclopedia that is outdated after only a few years.

The Encyclopedia as a physical book has become a chapter of history lessons. If I am looking for some information, I search via Wikipedia or enter my question directly into Google or YouTube. I think it works extremely well because it is successful even when I'm looking for a rare spare part for a household machine.

For today's kids, everything comes from the internet. Let me tell you a joke—which you may not even be able to laugh about: A two-year-old boy visits his mother in the hospital who has just given birth to another child. The boy asks: "Mom, where did you get this download"?

Back to the topic at hand: Today, almost all our children learn biology, physics, and above all math with the "simpleclub"[10] platform—at least in Germany.

And the music our kids listen to simply gets downloaded or streamed online—it comes from YouTube or from companies like Spotify. Many people don't even know what a CD is anymore, let alone a cassette.

Working in the Digital Age

Let's stay with the company Spotify for a moment. It was founded in 2006, so it is just 11 years old and has a market value of 3 billion euros today. Spotify has 200 million people as users in 78 countries.[11]

This rapidly growing company also has a very different organizational structure than traditional companies:

Classical Hierarchies: negative.
Usual boss roles: negative.
Normal project management: negative.

---

[9] https://en.wikipedia.org/wiki/Brockhaus_Enzyklopädie, accessed in April 2020.
[10] https://de.wikipedia.org/wiki/Simpleclub; https://www.youtube.com/channel/UCdQvwubOWGRB8JyoEC7lSbA, accessed in April 2020.
[11] The Verge: Spotify gets serious about podcasts with two acquisitions. https://www.theverge.com/2019/2/6/18213462/spotify-podcasts-gimlet-anchor-acquisition, accessed in April 2020.

Everything in this company is built on the principles of the so-called Agile methodology[12]; because otherwise you would be far too slow to develop new services and products.

According to the principles of Agile, there is no longer a classical division of labor. Working in largely autonomous teams has priority. Typical departments in project structures are nowhere to be found. Annual working time models often apply instead of fixed weekly working times.

> **The digital revolution does not stop at the smartphone. It has long since conquered the factory halls and is already conquering into all areas of our lives and work today.**

Here are a few examples:

There is no longer any reason to design public transport with buses and trams. Instead, we could imagine a small fully automatic eight-seater,[13] a kind of mountain gondola on four wheels that travels the route in 30 second intervals. One could then simply call them via an app and the next bus gondola with free seats stops in front of you. The same principle could be applied to cabs.

The fully automatic car will also allow people to make use of their car after the fourth glass of beer without any risk. How many of you think this will be very useful? I'm certainly looking forward to it.

What about queues at the registration office? This is completely unnecessary because you can schedule everyone for a specific appointment using an app. Soon, artificial intelligence systems will be able to manage such things much better than today's systems.

Why do we still have traditional retirement and nursing homes? Technologically speaking, we can reintegrate everything that is necessary to take care of the elderly into existing family structures. In many cases, mobile nursing and medical care are already the better choice. Medical monitoring, even at the level of intensive care, is already technically possible today in any private home that has a reasonable LTE or DSL connection.

An example of this ongoing transformation is a company based in Aachen, which operates over 40 ambulances in various cities in Germany via

---

[12] https://en.wikipedia.org/wiki/Agile_software_development, accessed in April 2020.

[13] Such fully automatic minibuses are currently being developed by an automotive supplier from southern Germany. https://www.automobil-industrie.vogel.de/zf-plant-joint-venture-mit-ego-mobile-a-609633/, accessed in April 2020.

tele-emergency doctors. This meets with great acceptance (as of 2020).[14] These tele-emergency doctors, based in the company's headquarter, are connected to the places where the accident happened via broadband communication. They have access to all relevant medical data of the accident victims and see the injuries in detail via video. They can control the admission to the best suited hospital depending on the injuries and in coordination with specialists much better than it was previously possible without that level of coordination. These systems are currently being introduced in many other regions of Germany.

The same company is also working intensively on the development of robots for the healthcare industry. This even includes the development of robots for dementia patients. As one can imagine, it is far less embarrassing to tell a robot than a human what it is, you may have forgotten. Certainly, the robot doesn't get angry because it has to repeat the same thing 50 times and is not emotionally disappointed when the memory performance of its counterpart diminishes.

Artificial Intelligence Is a Fundamental Gamechanger

Let's get this straight:

The digital revolution with artificial intelligence systems is spreading rapidly throughout the world.

> **Artificial intelligence does not stop at any cultural, national, or political boundaries.**

However, artificial intelligence will still be shaped by the cultural, national, or political beliefs of specific countries, as well as by the specific organizational conditions of where it was developed. For example, the detailed specifications of how autonomous vehicles should behave in traffic will differ (Maurer et al., 2015). Some will allow speeding. Others will demand a strict speed limit, even if it disrupts the flow of traffic.

The same thing happened with the Corona crisis in the year 2020. All governments applied AI-tools to track the spread of the disease in order to follow

---

[14] https://www.telenotarzt.de/en/, accessed in April 2020.

the flow of infections. But it is done under very different rules of implementation.[15]

At its core, we are dealing with an information revolution in which everything is connected to everything else. In Europe, this development is described by the term "4.0."

"Industry 4.0,"[16] for example, cannot be reduced to industry. All processes are connected to worldwide supply chains, both in terms of supply and customer service.

The same applies to "health 4.0." Here, too, the processes are increasingly interconnected worldwide. Today, an AI supported smartphone app is successful in the identification of skin cancer[17] and thus sustainably supports cancer prevention. Similarly, worldwide databases enable an up-to-date comparison of the relationship between prostate parameters and prostate cancer. Just as autonomous cars exchange their experiences about the roads they traverse on a daily basis.

Similar factors apply to "mobility 4.0." For example, I can already track the progress of my online order and it is already possible to deliver parcels with drones. I can use Google Maps to see the current traffic jams on every street in the world—seriously, all around the globe. New mobility chains are being planned which would make autonomous electric air taxi "Car2go" hubs possible. Booking would be handled through an app, and upon landing passengers could expect fully autonomous cars to be waiting to drop them off at an intercity express railway station, after which the car would continue along to its next scheduled destination. The search for a parking space would be a thing of the past.

Furthermore, completely new perspectives are opening up for energy supply. Decentralized energy supply with a local coupling of all types of energy is technically possible. Why should the price of electricity not be settled via a new approach, depending on the current availability of renewable energy in order to enable a faster energy turnaround? Why shouldn't all radiators in a house be controlled by an intelligent pump? This intelligent pump could replace the rather stupid thermostatic valve.[18] AI systems can be used to record

---

[15] How to Fight the Coronavirus with AU and Data Science (Feb. 15th,2020): https://towardsdatascience.com/how-to-fight-the-coronavirus-with-ai-and-data-science-b3b701f8a08a, accessed in April 2020.

[16] https://en.wikipedia.org/wiki/Industry_4.0, accessed in April 2020.

[17] James Vincent: Artificial intelligence can spot skin cancer as well as a trained doctor. Jan 26, 2017. https://www.theverge.com/2017/1/26/14396500/ai-skin-cancer-detection-stanford-university, accessed in April 2020.

[18] Wilo presenting geniax the heating pump on heating body (Feb. 19th, 2009). https://www.sbz-monteur.de/allgemein/wilo-praesentiert-geniax-die-heizungspumpe-am-heizkoerper, accessed in April 2020.

the energy requirements depending on the presence and absence of the residents. This means that a great deal of energy—up to 20%—can be saved.

And why shouldn't the fridge point out when it discovers mold? At the same time, I could use a speech system to tell the refrigerator all the items that I place in it, along with the expiration date. Some items will be scannable. The refrigerator will then remind me when I forget to consume something. Such systems already exist for blind people.

The message is not all of these things will happen exactly as described above. Nor is the message that all such systems will be centrally controlled, and that people lose their freedom. The message is rather: There are incredible opportunities for creativity, new scopes for design, and an immense variety of new possibilities to redesign life and work.

> **Will we use digitization and artificial intelligence for the good of the world, for the good of people, and for our own good?**

3

How Did Artificial Intelligence Come into Being and Where Do We Stand Today?

Contents
The Story of Inventions.. 22
Self-Learning Machines.. 23
Democratic Machine Control... 26
Quality Progress in Production.. 28
Robotic Teamwork.. 29

The dream of artificial intelligence actually began with the invention of the first programming language by Ada Lovelace[1] in 1820. Descartes'[2] world model was gaining momentum: the whole world was imagined as one big machine. This notion of the world as a giant machine runs through the entire history of technical development. Of course, such a model describes only a part of reality, as we know from many approaches to the theory of living systems[3] (Hanna, 1988). But initially, this evolution was characterized by a mechanistic world view. It was only much later that we began to understand that the behavior of organizations cannot be described by a mechanistic approach alone. For an appropriate representation of organizations, the characteristics of living systems are much more suitable.

---

[1] https://en.wikipedia.org/wiki/Ada_Lovelace, accessed in April 2020.
[2] https://en.wikipedia.org/wiki/René_Descartes, accessed in April 2020.
[3] https://en.wikipedia.org/wiki/OSTO_System_Model; https://henning4future.com/en/osto-system-model/, accessed in April 2020.

A key invention from the mechanistic approach is the first practical programming language, which finally achieved its breakthrough with the invention of the computer in the 1940s.

The Story of Inventions

By 1947, one of the first autopilot systems would be available to military aviators. Rudimentary systems like this one were the beginning of industrial automation, but not yet the beginning of artificial intelligence.

What we now recognize as the birth of artificial intelligence was a science conference at Dartmouth University in 1956.[4] Scientists discussed conceptual questions about computer languages, "self-improvement" of arithmetic operations, and the relationship between randomness and creativity. One focus was on "neural networks," which have since been used to simulate a mathematical model of human brain structures.

In the 1960s, automation technology followed with industrial robots, mobile robots, and the first chatbot "Eliza," a small software package with which scientists tried to simulate a psychological consultation. Back in 1992, this chatbot was already part of my compulsory university lecture "Computer Science in Mechanical Engineering" for first year students (Henning & Kutscha, 1994):

1. *How are you (program).*
2. I never have time for my friends (patient).
3. *Your friends?*
4. They're students.
5. *Could there be others?*
6. No, they talk about technology all the time.
7. *Can you give me an example?*
8. At the last party, all they did was talk about computers.
9. *And that is affecting you?*
10. I've already been dreaming about computers.
11. *What does this dream tell you?*
12. That I'm afraid.
13. *And that's why you came to me?*
14. Yeah, I need your help.

---

[4] https://en.wikipedia.org/wiki/Dartmouth_workshop, accessed in April 2020.

Of course, such dialogues don't require an AI machine yet. However, these limited discourses already demonstrate the tendency to perceive machines as living objects. My lawnmower robot is clearly stupid, but it's called "Moritz." And when I get home, I ask "How's Moritz"? Over the course of time, I have noticed that almost all people who own a robot as a lawnmower give it a name.

It seems obvious that it is very important whether we experience our machine counterparts as animals or persons rather than machines.

But all this is not yet artificial intelligence, even though we already occasionally label it as such.

We have spoken of "AI systems" at least since the first fully autonomous car completed a journey of over 212 km (132 miles) in 2005. Justifiably, because systems were used that autonomously mapped their environment and used the results to control performance on the road.

Self-Learning Machines

Since then, development has been going rapidly:

- Digital assistants that can be operated using natural speech such as Siri, Google Now, or Cortana have been in use worldwide since 2011.[5]
- The IBM Watson AI machine,[6] which beat its human competitors in a quiz test in 2011, has since been regarded as the industry standard for AI machine. Watson is widely used across the globe.
- In 2016, the AlphaGo AI machine won over the reigning human Go world champion. In 2017, it was in turn defeated by the newer AlphaGo Zero machine.

The development of the two AlphaGo AI machines is certainly a milestone that we need to take a closer look at conceptually.

Go traces its origin back 2500 years to China. The game has simple rules but an enormous amount of complexity. There are 2.57×10^{210} possible combinations. This is a number with 211 digits before the decimal point and is larger than the number of currently known atoms in the universe.

---

[5] Mona Bushnell: AI Faceoff: Siri vs. Cortana vs. Google Assistant vs. Alexa (June 29th, 2018). https://www.businessnewsdaily.com/10315-siri-cortana-google-assistant-amazon-alexa-face-off.html, accessed in April 2020.

[6] https://en.wikipedia.org/wiki/Watson_(computer), accessed in April 2020.

For comparison: The complexity of the game Go is 10^{100} times greater than the complexity of chess. However, from today's point of view, it is no longer a feat to build an AI machine that wins the chess world championship.

At its core, the AlphaGo machine consists of a combination of data-based learning and reinforcement learning, i.e. a combination of "frontal teaching" and learning by trial and error.

A so-called deep neural network was programmed, which consists of 12 layers. Each of these layers contains millions of neuron-like elements. The structure of each individual element is similar to that of a living being—a structure that has been very well researched. The machine was then trained by giving it access to a library of 30 million moves made by human players. Using this information, it was able to predict approximately 60% of the moves made by humans in previous games.

But that's not enough to win the game. "Frontal teaching" (teach-In) is—as in school—good, but not enough. To develop its own "non-human" strategies, the AlphaGo machine played thousands of games against itself. The learning strategy used is the AI machine DeepMind[7] from Google.

In this way, the AlphaGo machine trained itself. i.e. AlphaGo has undergone a kind of self-organized driving school. Finally, in March 2016,[8] it won four out of five games against the then reigning world champion Lee Sedol. In its winning run, the AlphaGo machine deployed moves that no human had ever made before.

This seemed to be a dramatic milestone in the AI development. But then a big surprise happened.[9]

> **AlphaGo Zero beats AlphaGo in 2017, which has beaten the world champion in 2016.**

Only one year later, on October 19th, 2017, a new machine—AlphaGo Zero—beat its just one-year-old colleague. In contrast with AlphaGo, only the rules of Go were given to this new machine. The "DNA" of the game alone—its rules—were enough for it to beat its predecessor.

Using "self-study," this machine gathered the necessary information from its environment, trained itself, and showed that self-organized AI systems that

---

[7] https://en.wikipedia.org/wiki/DeepMind, accessed in April 2020.
[8] Metz, Cade: In two moves, AlphaGo und Lee Sedol redefined the future. In: wired.com. https://www.wired.com/2016/03/two-moves-alphago-lee-sedol-redefined-future/, accessed in April 2020.
[9] https://en.wikipedia.org/wiki/AlphaGo_Zero, accessed in April 2020.

know nothing but the rules and stick to them can be superior to AI systems that work with a mix of teach-in and reinforcement learning. In AlphaGo Zero, human inputs are nonexistent beyond providing the rules of the game.

It is likely that in the future the same dispute will take place in artificial intelligence as it already does education, where traditionalist teaching methodologies are pitted against more modern approaches informed by developmental psychology: teachers as the sole source of knowledge flowing down to students versus teachers as moderators, enabling the creation of knowledge by facilitating student-directed discussion and employing online tools like simpleclub. But more of that later.

Well, where do we stand with the implementation of these concepts in practice? In the following pages, we will look at a few examples that illustrate the current situation and future outlooks.

> **Even weak artificial intelligence shows traces of intelligence.**

Let's start with an example of the so-called weak artificial intelligence.[10] In this case, intelligence is created by connecting independently acting software agents, i.e. connecting simple, independent, computing units. In order to understand this "character" of intelligence, let's consider the following pattern of gray scales in Fig. 3.1.

One can't make anything out at first glance. Only when one focuses on the interactions between the elements and pays less attention to the individual elements, the image of a face emerges—in this case the face of the author of this book. One can achieve this by either taking off your glasses, squinting, or placing the book in a distant corner of the room.

The information obtained from the interactions of the elements triggers another process of interpreting reality. We look for patterns, not for details.[11]

> **In highly complex systems with high dynamics, the importance of the interactions between the system elements is more important than the analysis of the individual properties of each element.**

---

[10] https://en.wikipedia.org/wiki/Weak_AI, accessed in April 2020.
[11] The combination of complexity and dynamics is called Dynaxity. https://en.wikipedia.org/wiki/Dynaxity. https://henning4future.com/en/dynaxity/, accessed in April 2020.

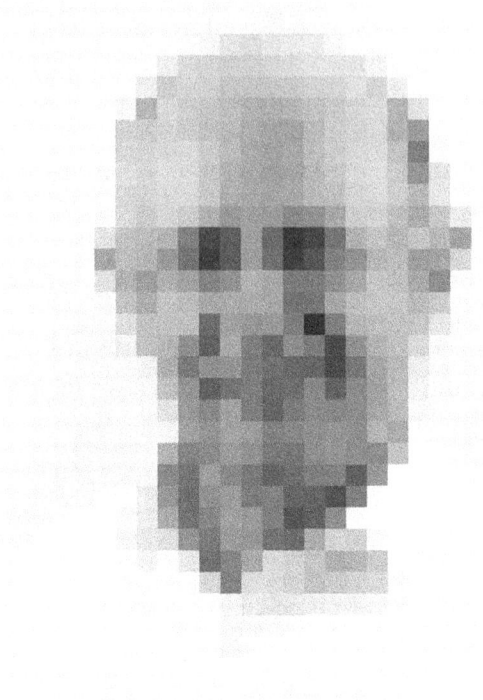

Fig. 3.1 Interactions are more important than the properties of the elements

Democratic Machine Control

This approach was first carried out in the Cybernetics Lab of the RWTH Aachen University in 2017 with an industrial knitting machine. All control devices (SPS controls, etc.) were removed.[12] All that remained were sensors and actuators through which to influence the machine. Then, the machine was equipped with 200 software agents, i.e. independent computers.

The classic tasks associated with the business of knitting were conventionally programmed and assigned to the software agents. There were agents for order management, quality monitoring, resource management, maintenance, communication, monitoring, perception, and voting.

---

[12] Abbas, Bahoz: Distributed Multi-Agents for Planning and Control of Production Environments Based on the Separation and Division of Powers. Dissertation. Aachen 2018.

Votings? Voting for what? Do software agents elect their bosses? Yes, in this experiment, they did exactly that in fixed terms.

To achieve this, a "social system" was developed for all 200 software agents, which was based on the democracy model of Germany, i.e. with a strict separation of the legislative, executive, and judicial branches, as well as layering according to the federal, state, and municipal structures. The top level regulated the responsibilities, the middle level production planning and control, and the lower level plant control. The horizontal separation of powers was supplemented by a vertical separation of powers.

The legislatures are embodied in the elections made by the software agents to determine their structure or the order management, which in turn determines what is to be done when and what is not. The judiciary is embodied in the way the software agents monitor production. Equipped with these features, the machine learns only through requirements, agreements, controls, but always according to democratic principles.

At the beginning, I regarded the experiment with a maximum level of skepticism; only to realize the following in the end:

The reliability of a machine controlled this way is significantly higher than that of conventional machines. To check this, the scientists switched off a randomly selected software agent 200 times—only to see its colleagues fully compensate for the shutdown agent within an average of 0.8 s (with low dispersion). The other "software colleagues" quickly took over the shutdown agent's tasks without interrupting the functioning of the knitting machine.

The second astonishing finding is that the start-up process was self-optimized without human intervention. This is an enormous improvement because the start-up of such machines normally requires a considerable amount of human intervention.

> A "democratized" machine control with network intelligence (weak artificial intelligence) already has a considerable advantage over the conventional centrally organized machine control.

Now, one could take this experiment even a step further: What would happen if we used 400 software agents and introduced a two-shift operation? And in addition, we could equip all 400 software agents with AI machines (for example, an IBM Watson machine) in addition to their programmed mandatory tasks. Then, we could let one shift work by the book, while the other 200 software agents use their Watson machines to monitor their colleagues' work, and after a few hours the two groups switch.

What would happen? I talked about this for some time with Sabina Jeschke,[13] one of the leading AI experts, and we agreed: It could indeed happen that the 400 software agents gradually remove themselves from their prescribed goals and tasks and, for example, form a software union through which new goals and task assignments are negotiated. It remains to be seen how this will affect product quality.

Here something like the Götterdämmerung of the central control of machines becomes apparent. This could mark the beginning of an era of democratization of machine controls.

The example of the knitting machine also shows the difference between weak and strong artificial intelligence. It becomes exciting to imagine the possibilities once machines can independently decide on their own goals and procedures to meet those goals.

Quality Progress in Production

The following example will of the repeat accuracy of welding seams. It shows the potential of applying strong artificial intelligence. The background: In mass production, the same welding seams must often be made again and again. But welding is a process that can only be reproduced with difficulty. In order to achieve this, you create extensive databases and try to analyze them again and again. One uses data mining methods, for example, in order to achieve improvements. If one achieves 60% repeat accuracy in relation to the desired outcome, one would be quite pleased. The remaining 40% is left to the experience of skilled workers who make corrections to the equipment.

If the question was whether AI methods could improve repeatability; the answer was "Yes," by about 50%. The obvious question would be, how so?

First, an idea was brought up from the gaming world, namely the game Super Mario,[14] in which a player has to overcome different flying or running opponents over and over again. The player can then—as with most games—progress to different "levels."

For this game, there exists a Super Mario AI machine that slowly upgrades itself to the higher levels, using algorithms in addition to neural networks that consider the specific task structure. The core of the approach is that the AI machine makes new predictions of what could happen next after each step and then decides.

---

[13] https://en.wikipedia.org/wiki/Sabina_Jeschke, accessed in April 2020.
[14] https://en.wikipedia.org/wiki/Super_Mario, accessed in April 2020.

This principle was then applied to the problem of repeatability of industrial welds and has since been implemented at the company in question. First, the AI machine was provided with all available historical data and then trained to use this information to make predictions. One has to imagine that despite the large amount of data, the database of welding surfaces is hardly uniform, with many jagged inconstancies, valleys, and peaks, all stemming from the inherent difficulties of working with molten metal. Nevertheless, the result was a 25% increase in repeatability.

Now, every welder knows that the quality of a weld seam also strongly depends on environmental conditions such as temperature or humidity. Therefore, detailed weather data (pressure, humidity, temperature, wind) from recent years were also made available to the AI system. Unsurprisingly, not one person can ever be able to process and evaluate so many variables at once. An AI machine *can*. The learning algorithm may be quite simple, but in combination with the huge amount of data available to the algorithm, significant improvements are possible. In this case from 60 to about 90%.

> In many "small" applications, artificial intelligence systems will produce considerable improvements in quality.

Robotic Teamwork

The breakthroughs of robotic teamwork in production, logistics, and assembly are well known.

The worldwide RoboCup competitions are a testament to these breakthroughs. In the RoboCup Logistics League,[15] autonomous robots are given logistical and assembly tasks that they have to perform as a robotic team—in competition against other teams. The robots move autonomously, similar to a motor scooter track at a fair (Fig. 3.2).

The team of RWTH Aachen University won the World Championship title four times between 2014 and 2017. What were the success factors?

The individual robot acts autonomously as a "super-agent." The system consisted of five robots and was operating radically decentralized, displaying the following characteristics:

---

[15] http://www.robocup-logistics.org/, accessed in April 2020.

Fig. 3.2 RoboCup world championship 2017 (Courtesy of the Cybernetics Lab of RWTH Aachen University)

- There is no central control. We've already seen a similar development with the knitting machine and its democratic control system.
- The robots have no hard-coded components. All components can change in their behavior and their structural parameters.
- An intensive sharing of all information takes place. The robots are all completely transparent in their behavior for the benefit of their "colleagues."
- Decisions are made cooperatively. There's no robot that has the last word.
- While working, it can happen that a robot revises its behavior strategy and, while doing so, restricts its activities strongly or completely. It's taking a creative break, so to speak. And the robot itself decides when to intervene again.

The decentralized control strategy marks a turning point for our understanding of technical controls. Generations of engineers have been trained to believe that this is not possible to achieve without a central control instance. As complexity increases, this new concept is becoming increasingly important, because the variety of decision-making options is much greater in decentralized structures. The decentralized control strategy keeps the system from causing conflicts in its hierarchical structure. This leads to a totally different

Fig. 3.3 Autonomous pallet control (Courtesy of Torwegge Intralogistics GmbH & Co KG)

kind of factory organization. And indeed, the control strategy of the RoboCup robots is already used in industrial practice (Fig. 3.3).

What is so special about this concept is the speed with which it can be put into practice. Within 6 months, these principles got converted into an AI-controlled autonomous vehicle for pallet control.

Central control can be omitted with clever AI systems.

Perhaps this approach also contains valuable lessons about how people might better organize themselves to maximize their distribution of power and responsibility. Of course, this is hard to imagine in humans. Nevertheless, from the successes of the RoboCup robots one might apply the following concepts to cooperation between humans:

- A team could develop its way of working without direction from a team leader.
- The behavior of the team members is not determined by fixed roles but adapts to the respective task and team situation.

- There are no secrets within the team. An intensive sharing of all information takes place.
- Decisions are made cooperatively. There is no single person who has the last word.
- Whoever needs a break decides that for themselves.

If one considers newer methods for product development—such as the principle of the Agile methodology[16]—some similarities are noticeable. Perhaps these approaches are frontrunners of a new kind of interaction between intelligent machines and humans.

Some indicators point into that direction: On factory floors, it is now possible to gradually dismantle the protective fences separating robot from humans. The interaction of humans and robots can be designed in such a way that the two do not collide with each other. Robots and humans can work together in a confined space during assembly work. Like two people, so to speak.

Another indicator is the replacement of the teach-in phase for robots capable of self-learning. As early as 2015, it was shown that a humanoid robot could learn to walk by itself without being explicitly taught how to do so. Such a robot is a simplified reproduction of the human musculoskeletal system. You let the robot have a go at walking—like launching a horse.[17]

In the beginning, the robot falls down all the time. After an hour it can walk a few steps. You allow the experiment to continue overnight. The final result: the robot will eventually learn to walk perfectly.

With the robot learning to walk we see the same process played out in the physical world as we did with the AlphaGo AI machine. The extensive teach-in that traditionally needs to be done to get such systems "up and running" can be replaced by learning through trial and error. The benefit being that, in contrast to humans, the machine does not get frustrated by many failed attempts.

> **In many cases, AI machines can replace teach-in with "learning by doing."**

AI systems will thus penetrate all areas of our life and work. In some places it will happen surprisingly fast and disruptive—in other places it will take decades until a functional technology is in place.

---

[16] https://en.wikipedia.org/wiki/Agile_software_development#Agile_management, accessed in April 2020.

[17] International Conference on Intelligent Autonomous Systems, Istanbul 2015. Siehe auch: TU Delft robot Leo learns to walk. In: https://www.youtube.com/watch?v=SBf5-eF-EIw, accessed in April 2020.

4

Can Machines Have Their Own Consciousness?

Contents

Intelligent Animals	34
Strong Artificial Intelligence	35
Machines with Ego-Consciousness?	37
Pedagogical Procedures of AI	38
What's Next?	39

Against the background of the experiential characteristics of artificial intelligence systems, the question arises: Can such systems develop their own consciousness above a certain performance level. The short answer: Yes, they can! But what is consciousness and how is it defined?

Almost all scientific fields have their own model world for the definition and description of consciousness. For the interaction of artificial intelligence systems and humans, it seems important to consider the perspective of how such systems are experienced by humans. Psychological, neuroscientific, and scientific approaches provide a good starting point for understanding these experiences.

Fundamentally, it's about how a system processes information and then reacts to it. As long as a computer system "only" receives information and processes it according to the "rules" of its developers in a way that the developer can predict which results will be produced as a "reaction" of the system, it remains an essentially deterministic process.

Even, if a system is able to create its own models of the environment from the observed environment and to generate "reactions" from this, it is still not very exciting.

However, it starts to get exciting when such a system begins to develop further its own models and produces conclusions and behaviors that the creator can no longer control and/or expect. At this point, we can finally speak of *operative* consciousness. Operative consciousness is a phenomenon that we are very familiar with in animals.

Intelligent Animals

Ravens are able to invent creative solutions from their wealth of experience, i.e. they create solutions that no one has taught them. This results in ravens being able to solve unknown tasks by tapping into their "stock" of experience. Their intelligence can be summarized in three categories:[1]

- They understand their environment. For example, they can throw stones into water in order to trigger movements that bring floating food closer to them.
- They are successful craftsmen. For example, they let cars drive over nuts and then wait until the traffic light turns red to eat the cracked nuts.
- They can plan. In one experiment, ravens were placed into an empty room without food for 1 day, and a different room with plenty of food the next. After a few days, both rooms were made available at the same time. The ravens planned to ration their provisions accordingly and took part of the food into the empty room. Furthermore, ravens are able to share their experiences with others and warn of situations that could be dangerous.

Already ravens have a first stage of consciousness.

This low level of independence is also the lowest level of consciousness. The associated intelligence is correspondingly higher in more highly developed animals. It is no longer controversial that rhesus monkeys are capable of reflecting on their own environment, i.e. they have their own "world view," which they are constantly developing, even if they are not observing anything.

---

[1] Ravens are Intelligent Opportunists. https://forestsociety.org/something-wild/ravens-are-intelligent-opportunists, accessed in April 2020.

This does not necessarily require contact with the environment. One can learn from indirectly observing, using these observations as if they actually came from direct interactions with the environment. The result of the reflection is therefore used again as input for another reflection. Thus, one comes step by step to an "imagined world view." And this idea of the world then becomes an essential basis with which to plan and implement actions.

When you ask a rider if their horse has a consciousness of its own, the answer is always "yes." He or she experiences the horse as a counterpart with its own consciousness. A rider also attributes feelings to their horse and experiences it as a living being with emotional states and expressions.

> Animals and humans can thus stand in intelligent exchange and form an intelligent partnership.

If we now address the question of whether systems of artificial intelligence have their own consciousness, we can take the described intelligence of animals as a starting point. From this understanding of consciousness, one can explore the question of how consciousness can form in systems of artificial intelligence.

Strong Artificial Intelligence

In these AI systems, a distinction is made between weak and strong artificial intelligence.[2] To answer the question of AI machine consciousness, it is necessary to examine those systems that are capable of forming strong artificial intelligence. But what exactly *is* strong artificial intelligence?

Systems of strong artificial intelligence are characterized by their ability to develop creatively and produce behaviors the developers could not program, design, or even imagine. These are systems that can think and reflect about their own state, so they know who they are and what they are for. Up to the point where such systems can also know from themselves when it is time to switch off, i.e. to bring about their own "death."

Whether it will ever be possible to build systems that reach the "level" of human consciousness is controversial. But there are at least valid arguments to the effect that this cannot be ruled out in the long term. Authors like Stephen

---

[2] Harlon Moss: Difference Between Strong AI and Weak AI. https://www.difference.wiki/strong-ai-vs-weak-ai/, accessed in April 2020.

Hawking[3] think this is possible and warn against it. Others like Markus Gabriel (Gabriel, 2018) consider it impossible.

Today, however, it is already the case that artificial intelligence systems are more intelligent in a certain segment of human's ability and act more intelligently than humans.

That this is possible is related to the fact that such systems can connect with each other and link their data simultaneously and worldwide. One example of this are globally connected travel systems that allow travelers to book hotels, taxis, or make flight reservations from anywhere in the world, right down to the last detail. In this way, a "Big Data" world is created that we humans are not capable of carrying around ourselves, either as individuals or in direct communication with others. This world of data is not only reflected in the global internet, but also in a variety of Global Area Networks.[4] The internet is nothing more than one of the many Global Area Networks, some of which are interlinked, but some of which are closed, company intra-nets.

On the other hand, the learning processes of AI systems are relatively easy to describe and not as complex as the structures of our brain. But in combination with the massive pools of data—the so-called data lakes—this is enough to become an intelligent counterpart for us. We experience the AI system as a counterpart with its own consciousness.

> **The combination of vast amounts of data and a simple construction of intelligence makes the special performance of AI systems possible.**

Let's get this straight: Consciousness is an integral part of systems that develop their own intelligence to a significant extent. In the current scientific discussion, there is an exciting approach for differentiating levels of consciousness. The AI expert Sabina Jeschke and the Linguists Riccardo Manzotti talk about the following:[5,6]

---

[3] Stephen Hawking: Automation and AI Are Going to Decimate Middle Class Jobs. Dec. second 2016. https://futurism.com/stephen-hawking-automation-and-ai-is-going-to-decimate-middle-class-jobs, accessed in April 2020.

[4] https://de.wikipedia.org/wiki/Global_Area_Network; https://www.techopedia.com/definition/7368/global-area-network-gan, accessed in April 2020.

[5] R. Manzotti and S. Jeschke, 'A causal foundation for consciousness in biological and artificial agents (Cognitive Systems Research)', Cognitive Systems Research, no. 40, pp. 172–185, Dec. 2016. https://www.sciencedirect.com/science/article/pii/S1389041716000024?via%3Dihub#!, accessed in April 2020.

[6] R. Manzotti and S. Jeschke, 'From the Perspective of Artificial Intelligence: A New Approach to the Nature of Consciousness', International Journal of Advanced Research in Artificial Intelligence (IJARAI), vol. 3, no. 12, Dec. 2014. https://www.researchgate.net/publication/287595546_From_the_Perspective_of_Artificial_Intelligence_A_New_Approach_to_the_Nature_of_Consciousness, accessed in April 2020.

- Thought Consciousness (A-consciousness): This area of consciousness includes the ability to develop models of the perceived environment and to reflect on them, i.e. including to monitor and control one's own perception. This model focuses on the consciousness of thoughts.
- Phenomenological consciousness (P-consciousness): This model focuses on the subjective experience of pain, temperature, or colors.
- Self-awareness: This is about the ability to differentiate oneself as a person from other people and to recognize and develop one's own identity—the ego-consciousness (Jung, 1995).

Machines with Ego-Consciousness?

For AI systems it is relatively easy to reproduce basic forms of human thought, i.e. to develop A-consciousness. This is "only" about the reproduction of partial structures of information processing in the human brain. This succeeds quite well with neuronal network feedback structures (deep learning).

P-consciousness is much more difficult because the intelligent systems for pain detection and processing in the human body are not only located in the brain. To build such "distributed" intelligent systems is far more complex. Here the analogy to locally and logically distributed network structures, which record and evaluate a common perception of a process, is obvious.

This also includes an intelligent structure of the "Internet of Things," i.e. a combination of intelligent AI systems that directly measure physical values such as pressure, temperature, and air composition and independently develop a model of the environment based on these measures. The AI system in these cases is inseparably embedded in the object.

The question of the AI realization of the third type of consciousness, self-awareness, remains unanswered. Whether AI systems can 1 day gain ego-consciousness in the sense of a person who distinguishes themself from other people is controversial in the literature. This is a very important issue which will certainly remain unresolved for a long time to come.

In my opinion, AI systems will never confront us as persons with their own ego-consciousness. But as I said, there are different opinions.

Let us return to the question of how systems of artificial intelligence develop a thought consciousness. That alone is dramatic enough.

How do systems of the so-called strong artificial intelligence learn? Let us approach this question in three steps.

Pedagogical Procedures of AI

First, such systems learn through a purely data-based methodology. In other words, they learn through observation as well as explanation. They collect data, and they are taught certain specific lessons. Design-engineers or programmers enter rules or certain schemas relevant for solving problems the system in question is designed to solve.

Such procedures are called "teach-in" procedures. In practice, this is nothing more than "frontal teaching." The systems are given tasks that correspond to the skills the developers have provided them with, and they are assessed accordingly on their performance in these tasks.

From pedagogy we know that such learning systems, which consist only of frontal teaching, are not very efficient.

> **Teach-in, comparable to frontal teaching, brings little intelligence to AI systems.**

The second step considers that many approaches have emerged in the education sector based on the principle of "learning by doing" (trial and error).

Take the Montessori approach,[7] for example, which is based on the idea that a child should discover what they *could* learn and be offered corresponding learning spaces for different subjects. In a learning room for mathematics, the child would be introduced to counting and arithmetic via a variety of materials.

It has been shown that learning by trial and error involves certain risks, but also has a very lasting effect.

Artificial intelligence systems also learn with this principle. It is known as reinforcement learning. The method uses the result obtained—the output—as input again, just as if it was an external input. And this feedback loop creates the feedback circuit of trial and error. A solution that goes in the right direction is rewarded.

From chaos theory we know that even simple multiplying feedback algorithms lead to chaotic, unpredictable results. The simplest chaotic system is a double pendulum, where the course of the pendulum already has a chaotic non-reproducible pattern.[8] The elementary feedback system with chaotic

---

[7] https://en.wikipedia.org/wiki/Montessori_education, accessed in April 2020.
[8] https://en.wikipedia.org/wiki/Double_pendulum, accessed in April 2020.

output is the so-called Verhulst's growth equation, from which chaotic patterns, called fractals, can arise (Henning, 1993).

The third step consists of combining teach-in procedures with reinforcement learning in artificial intelligence systems and linking them to a large database.

> **Teach-in plus learning via trial and error plus huge amounts of data have made the breakthrough of artificial intelligence possible.**

The concept was developed more than 50 years ago. But only today has it become reality, because the amount of available data and the performance of computers have grown exponentially and will continue to grow. What 50 years ago filled an entire bookshelf with punched cards as data carriers can fit on a small SD card today.

Today's computer technology is based on transistors as basic components. It is predictable that there will eventually be limits, simply because of the growing energy consumption of AI systems.[9]

What's Next?

Nevertheless, there is no foreseeable limit to the development of computer performance. This is because a new dimension opened up: Thomas Schimmel[10] introduced a new "transistor" in 2018. This single atom transistor consumes 10,000 times less energy than the transistors of today's silicon technology. If one considers that today's worldwide information processing done on computers accounts for about 10% of the electrical energy produced in industrialized countries, it becomes clear that a new era of computer technology is foreseeable.[11] Today, AI development is driven by new algorithms. The

---

[9] April Glaser: Artificial Intelligence Can't Think Without Polluting (Sept. 20th, 2019). https://slate.com/technology/2019/09/artificial-intelligence-climate-change-carbon-emissions-roy-schwartz.html, accessed in April 2020.

[10] Karlsruhe Institute of Technology: The world's smallest transistor switches current with a single atom in a solid electrolyte. https://www.nanowerk.com/nanotechnology-news2/newsid=50895.php, accessed in April 2020.

[11] Sabina Jeschke: 3 Shadoes of AI – 5G and Quantum Computing setting the stage for next generation AI. Inaugural Lecture at the Technical University of Berlin (22.01.2020). Compare too: https://www.zew.de/en/das-zew/aktuelles/kuenstliche-intelligenz-traegt-die-vierte-industrielle-revolution/, accessed in April 2020.

upcoming 5G technology together with IoT near orbit satellite technologies opens a new dimension of (real time) connectivity. And the third push will be created by the upcoming quantum computing technology. This will lead us to High Performance Computing systems (Bianchini et al., 2020).

If we think of higher levels of consciousness, which, for example, are shaped by the difference between the consciousness of humans and that of animals, artificial intelligence still has a long way to go. In this respect, I am convinced that this will not be achieved in the next century.

However, this does not change the fact that artificial intelligence systems will be superior to humans in certain areas through the combination of clever algorithms and vast amounts of data. We humans are not capable of simultaneously processing as many globally available and retrievable data as this is possible in the world of artificial intelligence.

For these intelligent algorithms, countless research groups worldwide are working to improve the core of artificial intelligence. The most commonly used principle is that of the neural network. In this process, known as deep learning, the individual elements of an input are fed to a layer of nodes. Each individual node has the same structure as a nerve cell (see Fig. 1). Several of such layers are connected sequentially up to one output. This output is then used as input—again and again. These feedback loops are the key to enabling such networks to continue learning from their experiences without necessarily requiring new input from outside. This cybernetic principle is also known from the functioning of biological cell structures.

If you want to apply this in practice, such a simple structure is of course not enough. For example, to build a digital rear-view mirror for self-driving cars that can see what happens behind my car in traffic, you need many such layer structures in parallel that interact with each other and learn from each other with nested feedback loops.

Even then, nothing has been gained because the developer then has to design the "DNA" of such a structure and define the framework conditions that apply to the cooperation of the different neural networks. And that's often not easy.

There is still a long way to go; the bottlenecks are in the details, not in the basic structure. Like is so often the case with technological developments, it's like running a long-distance race. Considering a race over a length of 100 miles, the last mile of reaching the finish line will take particularly long; often just as long as the distance to that last mile mark.

Current developments show that learning based on human-generated decisions often leads to worse results than "pure" learning without human intervention. It is therefore possible that the interplay between machine learning and human learning will develop in a completely new way.

We are only at the very beginning of this exciting development, to which the younger generation has already adapted to a considerable extent, as we will see in the next chapter.

5

The Homo Zappiens

Contents

Global and Local at the Same Time	43
Homo Zappiens at School	46
New Expectations Towards the Employer	47
Parallel Worlds	48
Communication with Images, Icons, and Emojis	49
Warning	51

The digital transformation of our lives began a generation ago. Communication with short messages, parallel life across numerous social networks, and the handling of automatic devices containing artificial intelligence have become normal for the younger generation since childhood. A completely new understanding of relationships between people and objects has developed.

A new type of human being is emerging that receives, filters, and processes information globally and is nevertheless regionally established. At the same time, awareness of global relationships and regional homeland connectivity is growing.

Global and Local at the Same Time

It is becoming increasingly normal for us to be in globally distributed, virtual work environments. Work is increasingly defined by *what* we do and not *where* we go to do it. The Corona crisis of 2020 accelerated this trend, encouraging us to rethink the distribution between the home office and working in the offices and factories.

A completely new understanding of global relationships is emerging, driven largely by the networking of personal data throughout the processes of our daily lives.

> **Working time and place of work are arbitrary.**

This "global milieu" poses a great challenge to the entire educational process. The key to new learning processes becomes experience-oriented learning. Learning must be fun (again), or learning does not happen.

During the Corona crisis, several of my grandkids found it much more pleasant to work at home and occasionally attend interactive video sessions with their teachers than sitting in a classroom.

In addition, we must learn to unlearn, as some authors say. The sociologist and prominent thinker of system theory Niklas Luhmann writes (Lehmann, 2012):[1] "The main function of memory is therefore to forget, to prevent the self-locking of the system from the accumulated results of earlier observations." In 2006, Wim Veen summarized the contours of this changed understanding of the human being in the digitalized world in one term (Veen & Vrakking, 2006): Homo zappiens.

The rise of the global-regional "Homo zappiens" is not about the disappearance of "Homo sapiens," but rather about two views of Homo zappiens and Homo sapiens that explain what has changed in the younger generation during the digital age.

As children, they acquire Homo zappiens skills that earlier generations acquired in the later decades of their lives or not at all. However, this does not mean that this development is only beneficial. It raises a lot of questions. What are the consequences of excessive media consumption, for example? However, it will not be possible to stop this development. It is much more important to use and shape it in a positive way.

Recently, one of my grandkids needed tutoring. The tutor was having a hard time with it and so was my grandson. Until the tutor suggested that my

---

[1] https://en.wikipedia.org/wiki/Niklas_Luhmann, accessed in April 2020.

grandson should listen to music in one ear during the tutoring lesson did the lesson begin to yield results.

> Homo zappiens learn in both highly parallel and non-linear ways.

What are the important characteristics of Homo zappiens compared to Homo sapiens (Gabriel, 2018)? Homo sapiens have the power to control themselves. But that doesn't mean they automatically do the right thing. Homo zappiens are not to be understood as the opposite of Homo sapiens, but as a complement.

The Homo zappiens are accustomed to making decisions at high speed and are permanently multitasking—doing several things at the same time. Homo sapiens are slower and can concentrate on a single task for a long time.

Wim Veen's reasoning is: That's a good thing! The young generation has already adapted to a world characterized by information overload. The task of education is then to treat a six-year-old's ability to multitask as an already acquired competence. On this basis, the educational process should attempt to reinforce the ability to concentrate on a single thing for a longer period of time.

Homo zappiens doesn't start reading at the top left of a page anymore. "Icons first" applies. Already at preschool age, children learn to navigate the internet with icons and pictures, even if they can't read. Many three-year-old's have more experience using tablets than adults, calling up their apps and knowing exactly where to find their clips on YouTube.

Homo zappiens love jumpy, non-linear approaches and connect everything before they look at individual elements. This is a skill that is perfectly suited for mastering complex tasks. This corresponds exactly to the competence of recognizing complex patterns by "blurred looking" (see Fig. 3.1).

Homo zappiens way of working is heavily based on networking. Preferably, learning takes place via searches, games, and a lot of imagination. The separation of learning and playing is unknown to Homo zappiens and learning by listening outright annoying. The power of imagination is sometimes so big that life in virtual spaces—such as Pokémon[2]—becomes more of a "reality" than reality itself.

Some time ago, an 11-year-old son and his father were together with me in the mountains. The boy had little desire to go hiking. But the expectation that

---
[2] https://en.wikipedia.org/wiki/Pokémon, accessed in April 2020.

at a reservoir at an altitude of 8000 ft. he could find masses of particularly rare Pokémon Go characters gave him a great deal of energy. In the evening, he showed me how crowded my garden was with Pokémon characters. He has opened up the mountain world through virtual reality. Because he suspected a larger collection of Pokémon Go characters up on the mountain, he accepted the climb. He was impressed not only by the virtual reality but also by the beauty of the mountains. For him this is a second reality: Augmented Reality[3] is the norm in his childhood.

Wim Veen would argue that, again this is a good thing, because the generation of the digital age has already adapted to the requirements of a globally networked information age and developed strategies for dealing with it that are meaningful and necessary. But what has *not* followed development at all—according to the Dutch professor of education and technology Wim Veen—is the educational process in schools and universities. Education must be based on the skills that children already bring with them from their networked children's room and life on the internet.

Homo Zappiens at School

To illustrate this point, Wim Veen conducted a school experiment in which the goal was to learn how to interpret a poem. The sequence of several lessons was essential an inverted version of a conventional teaching approach:

First, the students were asked to research in small teams on the internet interpretations and comments that already exist on the poem. They were instructed to do so not only in Dutch, but also in German and in English. They were then to produce a comparative assessment of the poem's interpretations and present it in class via PowerPoint presentations. Mind you: At that time, they had not yet read the poem itself. After a detailed discussion on how good the interpretations of the poem were, the text of the poem was gradually read in different languages. Only at the very end of the teaching sequence did the students actually write a successful poem interpretation "monocausal" and "slowly." The conclusion of Wim Veen:

> The school system will have to be turned upside down. The problem is not the students. Rather, it is necessary to change the pedagogical procedure in such a way that Homo zappiens can develop further based on his already acquired competences.

---

[3] https://en.wikipedia.org/wiki/Augmented_reality, accessed in April 2020.

5 The Homo Zappiens

This picture is also confirmed in the studies on Generation Y (Millennials)[4] (born 1976–1998). This is what the Münsterland Employers' Association in Germany writes about Generation Y:[5]

- "Generation Y is always online, always available, always up to date. The smartphone is no longer put away and is essential for all occasions."
- "They are always connected, work in teams, whether virtual or real, and they like flat hierarchies. They are willing to work intensively, but of course they also have to divide their energies, hence the great need for leisure and family life."

New Expectations Towards the Employer

However, this also creates other expectations towards the employer:

- The ethical behavior of a company is considered more important than the level of salary. For example, it can become a decision criterion whether a company participates in CO_2 compensation measures.
- The desire to have children has clearly shifted back to earlier years. Combined with the regulation on parental leave for men, this creates a new relationship to careers.
- As a result, the need for part-time solutions is also growing for managers. This was a taboo topic for a long time but is already practiced by many companies today. There are also more and more solutions in which two people share a fulltime position for a management task.
- More flexibility is expected for home office solutions. This expectation goes beyond the scope of traditional collective agreements. It goes even further: employees can temporarily stay in another part of the world and work from there every day.
- The company is expected to have its own offers for childcare and to adjust working hours according to the needs of families. In the diocese of Aachen, for example, a company was honored that worked as a call center 24/7, i.e. around the clock. The company introduced the rule that parents can bring their children to work without prior notification and without justification.

---

[4] https://en.wikipedia.org/wiki/Millennials, accessed in April 2020.
[5] Employer Münsterland: Generation Y: New ideas about the world of work. www.arbeitgeber-muensterland.de/blog/generation-y-neue-vorstellungen-von-der-arbeitswelt, accessed in April 2020.

- Sustainability of products and production conditions have gained a high value. Management seminars[6] clearly show that the question of the meaning of products and production conditions is increasing. Due to the shortage of labor caused by declining populations in industrial countries, young people have many more options when they begin searching for jobs.
- Flexibility is now demanded much more by employees than it is by the employer. It used to be the other way around. An IT consulting company in Southern Germany with over 500 employees has therefore given employees the greatest possible freedom to divide their annual working hours as they wish.

Generation Z[7] (born since 1998) shows a further development. This generation was accustomed from birth to interact with the great complexity and uncertainty of data, abundantly available via the internet. Generation Z attaches great importance to performance and job security or self-determined entrepreneurship.

> **Generations Y and Z (born since 1976) want a different life: Family, career, health, friendships, the need for security are all part of the so-called work–life balance, which should be in balance.**

This creative power of Generations Y and Z is becoming stronger and stronger in the highly developed industrial countries, as young people are in short supply due to falling birth rates. This is expected to lead to a dramatic shortage of skilled workers (Jeschke et al., 2015).

Parallel Worlds

Let's take a look at the "scene" in which this generation moves. I am referring to a book by Robert Campe (Campe, 2017), 16 years old in 2017, who vividly describes the "whereabouts" of his generation on YouTube channels:

---

[6] Systemic Management. https://henning4future.com/en/syma-systemic-management/; https://www.syma-seminare.de/, accessed in April 2020.
[7] Luecturi.de, 14.01.2016: Digital Natives: The 4 Generation Z Challenges for Employers. https://www.lecturio.de/magazin/generation-z/, accessed in April 2020.

- Videos on current topics are best viewed on MrWissen2go.[8] 1.2 million people subscribed (2020).
- On Emrah[9] you can find tips and tricks for everything. Founded in 2015, the number of subscribers grew to over two million within one year but has remained almost constant since then (2019).
- BibisBeautyPalace[10] is an absolute must for many young girls. In 2020, 5.8 million subscribed.
- With eight million (2020) subscriptions, Freekickerz[11] is still the most popular YouTube channel.
- And for school, every professional student uses simpleclub.[12] The platform has almost three million subscribers (2020) and provides an excellent way to learn mathematics, for example.

Every month new YouTube channels appear, and existing ones disappear. The professional Homo zappiens serves them all in parallel. Linear television, where you watch whatever is on at the moment, is becoming less and less accepted. Wim Veen demonstrates this in an experiment with students where they watch several simple movies at once without any issues. They use the image patterns to predict what the future course will be. And then they change programs. You just watch what you want and when you want it—on YouTube, Netflix, the archives of TV programs, or other platforms.

> Being online on many (social) media channels simultaneously has become a normal part of life.

Communication with Images, Icons, and Emojis

Even more important, however, is the fact that the world of icons has developed into its own language. Every user of a smartphone will find a set of about 2000 emojis to replace letters, which can also be used to write. This also helps a lot in work with refugees. Many conversations are initially only possible with pictures or icons. A German design company[13] developed a special template with icons on two A4 pages to facilitate conversation.

---

[8] https://www.youtube.com/user/MrWissen2go, accessed in April 2020.
[9] https://www.youtube.com/user/CrazyEmri2, accessed in April 2020.
[10] https://en.wikipedia.org/wiki/Bianca_Heinicke, accessed in April 2020.
[11] https://de.wikipedia.org/wiki/Freekickerz, accessed in April 2020.
[12] https://de.wikipedia.org/wiki/Simpleclub, accessed in April 2020.
[13] http://amberpress.eu/buecher/icoon-first-help-refugees-welcome/, accessed in April 2020.

Fig. 5.1 You can talk with WhatsApp (Campe, 2017)

Another example: Robert Campe reports on a dialogue in which his mother writes him a longer message (Fig. 5.1) using emojis.

I use these lines again and again as a "reading test." With ten-year-old readers, that's no problem at all. They read such a sentence almost fluently, but with certain inaccuracies. It is interesting in this context that emojis can look very differently depending on the operating system or apps used. With WhatsApp on a Samsung smartphone, you can switch between two different emoji sets, some of which allow significantly different interpretations. Nevertheless, this does not seem to lead to misunderstandings.

Just as in the epoch before Gutenberg everything had to be explained in pictures because people could not read, today the world of pictures returns with icons and emojis.

Most older people cannot read this sentence. So far, I have experienced it only once that a 50-year-old could read this fluently.

This kind of communication changes communication in general: the images and mythical representations are coming back. That's no surprise, because we have learned to orient ourselves on the internet with many pictures and icons. However, images and myths are always more blurred and unclear than text.

> **The images and myths are coming back.**

Dear reader, perhaps you have also understood approximately what the icon sentence means. If you want to know more, the sentence is:

> I'm not picking you up in the car because I'm going out for coffee. Take the bus home and bring a cake from the shop at the train station.

Dealing with media has many new dimensions:

- Conferences are possible "anytime, anywhere."
- Writing 200 relevant e-mails with valuable content per day is possible. I have witnessed this firsthand.
- Direct communication without secretaries is possible.
- Writing with automatic source generation actually works. Yes, there are also first attempts that AI systems write books.[14]
- Informal networks are easily organized.
- Flash mob events are fast, effective, and feasible at a large scale.
- The library of management tools available to individuals and businesses is growing. Leading via a variety of social media tools becomes normal.
- I can "talk" via WhatsApp icons, etc.

Warning

On the other hand, the dangers of this development cannot be overlooked. One of the critics—Manfred Spitzer—strongly warns against the exaggerated use of social media (Spitzer, 2012). His main criticism is that this approach only superficially addresses the content. He argues that it diminishes one's own creative learning. However, this cannot be prevented by prohibiting social media. It is part of the responsibility of parents and teachers to show ways in which this risk can be reduced. This is exactly where Wim Veen's approaches have proven to be groundbreaking.

However, the risks in dealing with the digital possibilities are also considerable in practice. The following are some ironically meant "options" in the sense of the "logic of failure." In other words: How do I behave successfully wrong? Here are some paradoxical recommendations:

- Stop talking to your employees and communicate only via e-mail, WhatsApp, etc.
- Always use as many cc's as possible so that everyone stays informed what you are up to—take care of your information duty through cc.
- Only notify restructuring measures, dismissals, etc. electronically.
- Use bcc purposefully so that your information policy remains your secret.
- You can be reached 24/7 and react after 10 minutes at the latest to be efficient.

---

[14] Schäfer, Michael: Artificial Intelligence: First book of an algorithm published. In: computerbase.de. https://www.computerbase.de/2019-04/kuenstliche-intelligenz-erstes-buch-algorithmus/, accessed in April 2020.

- Keep your kids away from their smartphones. And then they watch you checking your phone while they eat.
- Never have a conversation without being active online at the same time.
- Delete your inbox after four weeks at the latest so that you are not reminded of any unfinished business.
- Avoid stopping by at your colleague's room if you can communicate with him/her via e-mail.
- Send out all relevant e-mails on Sunday afternoons so that everyone arrives prepared at their workplace on Monday morning.
- Avoid times when you are offline. Avoid times only taking care of yourself.

There is much to learn to find a good balance between Homo zappiens and Homo sapiens. Homo zappiens is no alternative to Homo sapiens. Rather, the term describes the expansion and change of human learning and behavioral processes in the digital transformation.

The many new dimensions of learning and communicating will further increase the complexity and dynamics. This will also lead to limits of mental resilience. It is about the art of making selecting the various new possibilities usable in a targeted way—like going shopping in a supermarket. You can choose from an enormous selection, and you will choose "your" supermarket.

In this way, completely new and diverse habits of learning, living, and working are emerging worldwide. Companies will adapt to the new expectations of young workers. Schools will fundamentally rethink their teaching and learning concepts and humans will learn to cope with the high amount of information and the parallelism of processes that can be overwhelming. They will also have to learn to find a new relationship between activity and leisure.

6

The Inverse Gutenberg Revolution

Contents
Social Bots.. 54
A Transparent Society.. 56
Ethical Boundaries... 57

We just learned that images and myths are finding their way back into our communication. We are already experiencing this in WhatsApp conversations. But the world of text messages has also taken on a different character in recent years. More and more people communicate with "picture messages" in which logical connections play a minor role. Rather, we see simple statements getting spread on the internet, via Instagram, Facebook, Twitter, or any such medium that has not even been invented yet.

There is more to that. Our mental model of communication is being changed sustainably. Pictures, short messages, and "impressions" are spread on a massive scale. In this way, behaviors and ideas can spread quickly around the world. This phenomenon is what is known as the "Meme" and goes back to Richard Dawkins (Dawkins, 1989).

That would not be so far-reaching in itself. What matters is that these "picture messages" elude rational discourse. The dynamics of the "Likes"—I like it or I don't like it—determines the dynamics of the distribution and influences my personal opinion. If it is targeted correctly, it will also be increasingly successful in shaping political opinions.

That in itself would not have very far-reaching effects. What matters is that these "picture messages" elude rational discourse. The dynamics of "Likes"—I like it or I don't like it—determines the dynamics of the distribution of such memes and influences my personal opinion. If it is targeted correctly, it will also be increasingly successful in shaping political opinions.

Social Bots

An essential dimension of this new way of communication is the following: For the creation of such short messages and pictures I do not necessarily need people who "create" these messages and pictures. This can also be done by automatic machines and/or intelligent AI systems.

Let's take a closer look at that through the example of the "Twitter world." Approximately 15% of all Twitter accounts worldwide are not operated by people, but are automatic accounts, the so-called social bots. The 15% amounts to about 48 million accounts.[1] Such a social bot is an—increasingly intelligent—computer program that replicates human behavior patterns in social media. These social bots are on the move on the internet without human intervention, in our case in the Twitter world.

A larger proportion of these 48 million accounts have "bad" intentions, whereby "bad" is a very relative term. These "bad" automated accounts range from the dissemination of right-wing extremist ideas to advertising for illegal and immoral processes and election campaign machines.[2]

The more intelligent of these social bots can also reproduce images of living people and let them talk. They can do this well enough that the viewer has no chance of distinguishing this simulated person from a real one.

Thus, the Star Trek fiction of replicants in two-dimensional space has become a reality in our world: People who talk and act like people but are no real persons. More precisely, it is the image of the three-dimensional person on the surface of the screen.

A well-known example of this was a group of students from a renowned American university who intervened in the election campaign with a replicant Barak Obama.[3] The lip movements of the synthetic Obama were generated by an AI machine with neural nets. The students then took sentences and phrases from Obama's speeches and used them to create lip movements. As a result,

---

[1] Erxleben, Christian: Social Bots on Twitter: 48 million accounts are not people. In: BASIC thinking https://www.basicthinking.de/blog/2017/03/17/social-bots-twitter/, accessed in April 2020.
[2] Abbas, Bahoz: Artificial Intelligence in Politics - The Way to Free Individual Opinion Formation in Social Media. Doctoral Lecture RWTH Aachen 11.09.2017
[3] IEEE Spectrum: AI Creates Fake Obama. In: spectrum.ieee.org (26.04.2019). https://spectrum.ieee.org/tech-talk/robotics/artificial-intelligence/ai-creates-fake-obama, accessed in April 2020.

the viewer has virtually no chance of identifying whether they are seeing the real Obama or the replicant Obama.

Such applications for creating videos with faces that look deceptively similar to real people are now downloadable as a "Deep Fake App"[4] for everyone from the internet.

Is that politically justifiable? Should that be forbidden? If so, how?

Three-dimensional holographic representations will follow. The same goes for films in which, for example, a replicant Obama walks around. But I don't think it's possible that one day a robot Obama will run around that I can't distinguish from a human anymore. To be on the safe side, I would have to touch his skin...

Another dimension is created by automatic accounts that move in the political scene. An example of such an intelligent social bot became known where an automated social bot had gotten involved into the right-wing radical scene. Consequently, its entire communication history was deleted. But the memory patterns that this social bot had already formed remained. His "world view" had become part of the bot's DNA.[5]

The social bot, robbed of its memory in this way, was then—with its "view of the world"—released back onto the net. The result was almost to be expected: Within a very short time, the social bot's comments were once again right-wing radical. It was shut down: The death penalty for intelligent machines, so to say, if these do not correspond to the desired world view.

This example shows the difference whether I "only" have a huge amount of data or whether I can do something with this data in combination with an intelligent AI machine.

Big Data alone delivers volumes of data in an increasingly shorter time. This creates complexity and dynamics, a considerable increase in the so-called dynaxity.[6] The term dynaxity is composed of the two terms complexity and dynamics and describes the combination of complexity and speed of change. The degree of dynaxity is divided into four zones: (1) static, (2) dynamic, (3) turbulent, and (4) chaotic. While traditional methods of perception and control are effective in zones 1 and 2, the transition to turbulent zone 3 represents a serious paradigm shift. One aspect of this is the enormous increase in available information, which is reflected in data dynamics.

---

[4] https://www.chip.de/downloads/Deepfakes-FakeApp_133452282.html, accessed in April 2020.

[5] Vincent, James: Twitter taught Microsoft's AI chatbot to be a racist asshole in less than a day. In: The Verge. https://www.theverge.com/2016/3/24/11297050/tay-microsoft-chatbot-racist, accessed in April 2020.

[6] https://en.wikipedia.org/wiki/Dynaxity, accessed in April 2020. https://henning4future.com/en/dynaxity/, accessed in April 2020.

A Transparent Society

This data dynamic creates a transparent society.[7] Much more information than ever before is becoming transparent, accessible to nearly everyone. This effect of the transparent society also has positive effects. For example, Google has succeeded in anonymously identifying all active mobile phones and smartphones and deriving movement data from them. But this data wouldn't have had much value alone.

Only the use of AI algorithms, which evaluate these data with given goals, results in a new dimension. In the case of Google, the traffic jam indicator of Google Maps was created, which functions by collecting the movement data of most mobile phones in the world. This makes the global traffic situation around the globe completely transparent. From the couch in my living room in Germany I can check what it would be like if I had to cross one of the bridges from Newark to New York City. I can see what alternatives would be possible at this moment. There is complete transparency of all the traffic jams taking place on roads somewhere in the world right now. Including Hong Kong. But not the traffic jams of Beijing and Shanghai, because the government is blocking access to this data. More about this later.

> At its core, the combination of big data and artificial intelligence creates the possibility of the predictable human being, a dimension that goes far beyond the transparent human being.

In the political sphere, the same procedures allow direct intervention in election campaigns. This also means that any regional and national election campaign can at the same time become an international election campaign if there is access to the global open network.

With this background it is no surprise that politics can be made with a Twitter account. One can win a vote with pictures, myths, or short messages that can consist of both true and fake news.

The worldwide dynamic of data that is available everywhere in combination with automatic accounts has already created a new culture around the globe: Highly efficient for useful processes such as traffic jam forecasting and highly dangerous for political stability and democracy.

It seems as if the significance of reason is diminishing. Just like before Gutenberg's time, there is communication via images, myths, and short news

---

[7] http://www.davidbrin.com/transparentsociety.html, accessed in April 2020.

that has a lasting impact on people's lives and behavior. I call this the Inverse Gutenberg Revolution.

> We are still far from being able to perceive, evaluate, and control the power of the instruments of artificial intelligence that are already available and in use today.

One could compare the development of artificial intelligence with the invention of atomic energy. At that time, the inventors had an idea about the opportunities and risks of this energy. They even spoke about it publicly. But it took decades until their warnings were taken seriously.

Ethical Boundaries

Today, we already see that artificial intelligence has become an extremely powerful weapon that is used in all areas of the world—from useful, practical application to new vehicle concepts, factories, and mobility concepts to the global influence, manipulation, and control of people.

Do we hear and understand that all this is much more radical than, for example, the upheaval from horse-drawn carriages to motor vehicles?

Do we still believe the AI experts who tell us it's all not as big a deal and it is all just computers and we've had them for over 70 years now?

In addition to redesigning our factories, our mobility systems, and our way of communicating, it's all about redesigning our social and political systems—worldwide. And that is where our democratic system is at stake. Let us shed some light on the dimension by using the example of face recognition through AI systems:

In the past, a computer system had to be loaded with a wide collection of faces via the teach-in process in order to enable it to identify faces on its own. This is a technique that police criminal investigation units have already been using for decades.

Today, an AI system only needs the rules to search for the "round part" and sort the Euclidean distances in space: "Measure all distances in the three-dimensional space and evaluate them at your own discretion." The amazing thing is: After a short time, the system can distinguish between male and female faces. Of course, the system first distinguishes between A and B and then needs to be instructed that A means "men" and B means "women."

With this system, the People's Republic of China now carries out the nations' education for 1.5 billion people. They even complement the use of facial recognition cameras, by monitoring payment behavior, for example. If you behave well, you get bonus points; if you behave badly, you get malus points. The underlying ethical standards—whatever they may be—are determined by the state.

In China's big cities, every person who crosses the street on the red light is now "caught" and gets penalized.

One of my colleagues told me about an actress who was penalized for such an offense, while she definitely hadn't crossed the street at that particular time. She complained and it turned out that at the time the camera had supposedly recognized her face on the street, a bus had passed through the pedestrian crossing in question. On the side of the bus at a height of 6 ft., a picture of this actress could be seen—about the size of her head. And that was all the camera needed to log a penalty in the system. Luckily, the situation was rectified, and the penalty was deleted.

According to our democratic understanding, most will argue that such a state system is irresponsible and inhumane and violates the fundamental right of human self-determination.

However, there is evidence of a surprising "side effect" of this system. Studies have shown that Chinese citizens think very positively about this system. One of the reasons is the hope of decline in corruption. But the system was shut down by officials (2019).[8]

Should the Chinese government have found a way to fight corruption sustainably? And if that turns out to be the case, how do we assess the value of "corruption prevention" versus "data protection"?

And would the containment of the spread of the coronavirus in China in 2020 have been possible to the extent we saw without a sophisticated, nationwide, AI-based digital control system?

If we think of a continent like Africa in this context, perhaps the fight against corruption could be a higher value than data protection—an ethical and social issue that is not easy to resolve.

We don't even have to wander into the distance to see the applications of facial recognition. Some time ago I lent my skiing pass (which contained a picture of myself) to my son. The next day, the pass was disabled. I found out

---

[8] Stephen Chen: Is China's corruption-busting AI system 'Zero Trust' being turned off for being too efficient? (Feb. fourth, 2019). https://www.scmp.com/news/china/science/article/2184857/chinas-corruption-busting-ai-system-zero-trust-being-turned-being, accessed in April 2020.

that all skiers in the widely distributed ski area were digitally scanned at all ski lift valley stations and that the facial recognition—despite ski goggles—made it possible to discover beyond doubt that the ski pass did not fit the owner. No one was asked when this system was introduced. Neither did I sign any consent for the automatic face detection.

There are numerous initiatives on the question of how to rank the values of artificial intelligence systems. The European Parliament intends to build a platform on which all current algorithms of artificial intelligence can be found. The initiative is to result in an ethics charter for artificial intelligence. In my opinion, we will need such ethical limits for the digital transformation with artificial intelligence.[9]

For example, the protection of individual opinion-forming would require automatic social bots that contain "ethical algorithms." Such a bot hunter would then have to use ethical criteria to search the internet for violations of ethical principles. In case of doubt, such "policemen on the internet" would form working groups worldwide and elude state legislation. It is therefore urgently necessary to develop an international agreement on AI ethics tests which could include, among other things, an assessment of the content according to ethical/moral principles, such as truthfulness, respect for human dignity, etc.[10]

How difficult this is can be seen in the discussions about the worldwide copyright law. On the one hand one tries to exercise control with the so-called upload filters. At the same time, it is precisely this control that allows a significant increase in power for those who operate the upload filters.[11] The filter algorithms that are used in this process make misjudgments, just like humans.

Similarly, problematic was ambivalent the development of the text generator GPT2 by OpenAI,[12] a non-profit research institute for artificial intelligence. The company pulled a version out of circulation that was too strongly influenced by external opinions.

---

[9] EU Commission plans comprehensive European initiative. https://ec.europa.eu/digital-single-market/en/artificial-intelligence, accessed in April 2020.

[10] Abbas, Bahoz: Artificial Intelligence in Politics - The Way to Free Individual Opinion Formation in Social Media. Doctoral lecture RWTH Aachen. 11.09.2017

[11] Kaube, Jürgen: Copyright and Uploadfilter - Große Hehler. In: Faz.net, 22.03.2019. https://www.faz.net/aktuell/feuilleton/medien/artikel-13-urheberrecht-und-uploadfilter-kommentar-16101425.html, accessed in April 2020.

[12] Whittaker, Zach: OpenAI built a text generator so good, it's considered too dangerous to release. In: techchrunch.com. https://techcrunch.com/2019/02/17/openai-text-generator-dangerous, accessed in April 2020.

Considering the many risks that go along with using artificial intelligence, we will have to discuss the ethics of AI systems with their own consciousness in more detail (cf. Chap. 12).

> **The Inverse Gutenberg Revolution links everything with everything and will lead to the fact that we will have to find and agree on a new—worldwide—order of values.**

The core of today's cultural revolution (the Inverse Gutenberg Revolution) consists of a development towards machines, devices, and service products which, through the combination of mass data, neural networks, and learning algorithms of artificial intelligence, have their own consciousness. They are in a lifelong learning process and may even decide to take themselves out of operation in time, in other words to "die."

7

The Age of Hybrid Intelligence

Contents
Technology–Organization–People.. 62
Driverless Trucks: Why Is It Taking So Long?... 63
Added Value of the Man–Machine Interaction... 65
Shadow Economy... 67

With the background of the Inverse Gutenberg Revolution, the question arises how the interaction of humans and intelligent machines can be shaped.

For decades, the paradigm of the HOT approach (First Human, Second Organization, Third Technology) applied to me (Henning, 2014):

- First people,
- then the organization,
- then the technology.

This is based on the hypothesis that the focus lies primarily on people and that the person responsible for the design of organizations and technology is the human being. In principle, this is still valid.

Obviously, our behavior plays an important role in the design of organizations and technology. It is astonishing that an organization such as the Office for the Protection of the Constitution of Bavaria/Germany[1] points out that

---
[1] Elsasser, Thomas: Dangers of industrial espionage. Annual Conference of Xenium AG. Munich 16.10.2015.

the core of any protective measure lies with the human being and after that with the organizational structures. Only then do developed technology systems makes sense.

However, the practice of technical innovation is different. The same phases can be observed with every new technology.

Technology–Organization–People

In the first phase we usually fall in love with new technologies. In this phase, the players believe that with this technology automatically everything will be fine. This has been the case with all automation waves in recent decades. Already at the end of the 1980s of the last century "Computer-Integrated Manufacturing" (CIM) was proclaimed and the deserted, fully automatic factory was expected shortly. A few years later, the concept "HCIM—human-centered computer-integrated factory" boomed (Brandt, 2003).

Then, people realize that some organizational changes are needed to make the whole thing work. Normally, special departments are then founded, for transformation, cultural change, or agility. This was also the case with the innovation wave "knowledge management" in the 1990s of the last century (Henning, Oertel, & Isenhardt, 2003). After it became apparent that the automatic knowledge databases did not solve the problem, a wave of organizational measures followed. Special departments for knowledge management were founded.

Only when it becomes clear that the organizational adaptation measures are not enough does one reflect on the human being and their significance in the system. In a large-scale administrative reform project in Germany, the best practice examples nationwide were made available online. In all those best practice cases, the connection was always made to a person at the center of the organization who could provide information from person to person.

Unfortunately, the practice usually proceeds in the following order: first technology, then organization, then people.

The development of the AI wave in recent years is no different. In 2009, one of my research teams drove over 3000 miles in traffic on the German Autobahn with coupled truck convoys of various types (Henning et al., 2009; Henning & Preuschoff, 2003). Four trucks were coupled at intervals of 33 ft., with a load of 160 t at 50 miles/h and only in the first truck was a driver actively steering (Fig. 7.1).

Fig. 7.1 Automatic truck convoys in flowing traffic in Germany in 2009 (Haberstroh, 2014)

Driverless Trucks: Why Is It Taking So Long?

In the example of truck convoys, there have been years of discussions on legal and transport policy principles that have resulted in several new research projects. Nevertheless, these projects didn't necessarily lead to new insights. In one, after 10 years of initial road tests and over 20,000 miles worth of trips carried out in traffic at a distance of 50 ft. between trucks,[2] it was found that the gap is too wide to achieve significant energy savings—a result that was already clear in 2009. For this reason, a distance of only 33 ft. was chosen at the time.

Everything around driverless trucks is technically solved and has been researched several times. Nevertheless, there are still a few organizational problems with case law and cross-border transport. This is nothing new: It often takes a little longer for finished technologies to impact everyday life.

---

[2] Joachim Becker: Dampers for the trucks on the string. In: Süddeutsche Zeitung.de, 21.05.2019. https://www.sueddeutsche.de/auto/autonomes-fahren-platooning-test-1.4444322, accessed in April 2020.

And then there is also the human factor. In the year 2017, a major debate broke out that by the year 2030 up to 4.4 million of the expected 6.4 million truck drivers in Europe and the USA will become unnecessary.[3] At the same time, there is already a shortage of 100,000 bus drivers in Germany.

But will truck drivers really become unnecessary? Could that really happen within the next 8 years? No, that's not going to work! Why? One must consider the question of acceptance in society. How long do we need until not only engineers and computer scientists, but we as a society have the confidence that these 40-ton trucks are allowed to be all driverless on the roads at 50 miles per hour?

How long will it take before there are legal regulations to make this scenario a reality? According to all estimates this type of traffic will cause significantly fewer fatalities than conventional traffic. Perhaps one day it will be possible to prove that the operation of fully automatic cars on the roads is safer than the operation of cars with human drivers.

We must also distinguish between various modes of transportation: Is the truck driving on the motorway or on a country road? Does it move in a logistics center or in the middle of the city center? Of course, it is relatively easy for the driver to get out of the truck at the gate of a logistics center and the rest of the slow journey is fully automated with an AI system. This would include shunting backwards to the ramp. But if we think of a snow-covered country road where the edge markings aren't visible anymore, combined with poorly visible pedestrians on the road, then it becomes much more difficult without a human driver.

Even if fully automatic driving of trucks on motorways is permitted (this would already be technically possible to a large extent today), would freight forwarders then allow freight to be transported by road without supervision? Isn't there a much more sensible way where the truck driver's cabin could contain office equipment? In this scenario, they could remotely carry out a desk job in their company and continue driving the truck himself after leaving the motorway.

And what about the fully automatic refueling? Of course, we can build a tank robot that opens the filler neck and refuels automatically. However, for this to happen, there must be a reliable, nationwide network of automatic tank robots that operates faultlessly along all possible routes.

---

[3] Mortsiefer, Hendrik: Robotic trucks threaten millions of jobs. In: Tagesspiegel.de, 31.05.2017 https://www.tagesspiegel.de/wirtschaft/autonomes-fahren-roboter-lkw-bedrohen-millionen-jobs/19871754.html, accessed in April 2020.

And how do you organize the protection of trucks in case of a burning tire? How does the AI system recognize that the tire is burning? How does the AI system decide if the truck is to be driven further so that the chassis does not catch fire? In this case, how does the AI system recognize whether the rubber blanket has burned down to such an extent that the truck can be stopped? How do you deploy emergency warnings after a breakdown that leads to a standstill on the side of the road?

There are also economic issues. The service life of a truck today is at least 8 years. Forwarding agencies calculate with this lifespan. They cannot afford to take all conventional trucks out of service at once and replace them with fully automated ones. Even if all the new trucks were only allowed to drive fully automatically, the changeover would take at least 8 years—under the assumption that we don't want to ruin the freight forwarders. Quite apart from the fact that, from a production point of view, it wouldn't be possible to do it any faster. Nobody wants to "quickly" build a truck factory and then close it down again after some years. That wouldn't pay off, and the trucks would then become far too expensive.

> **Conversions of system and vehicle technologies always take decades, even if one does not want to admit it.**

Well, the idea of driverless trucks for all traffic areas won't be realized all that fast. It makes much more sense to link human and AI system in a meaningful and efficient way. This is because the combination of the work of AI machines and that of human beings can be implemented much more quickly.

Added Value of the Man–Machine Interaction

For example, the weekly working hours of truck drivers could finally be reduced to a reasonable level. And the truck driver's workplace could be upgraded and made more attractive.

In most areas of vehicle and system engineering, there will be used a "hybrid intelligence" in which the capabilities of humans and AI machines are combined in a targeted manner. It is about building trust in human-centered AI.[4]

---

[4] There is a lot of discussion about this approach in the European Union. https://ec.europa.eu/digital-single-market/en/artificial-intelligence, accessed in April 2020.

At its core, this is not a new thought. The importance of human–computer interaction has always been the decisive factor in the implementation of new automation technologies.

But the relationship between people and AI machines and vice versa changes fundamentally once AI machines have their own consciousness. The result is a counterpart with whom communication takes place at eye level.

> **A new kind of partnership between the human being and the AI machine emerges.**

Humans often have a close partnership with animals. A rider and a horse are a "system" with such hybrid intelligence, in which the horse does not work like a machine, but like an intelligent counterpart with its own will and its own decision-making capability. Only if the symbiosis between human and horse is successfully will the team win a competition.

From my experience as a horse-drawn carriage driver I know that the interaction becomes even more complex when you have four horses in front of you instead of one. The four horses interact with each other as a team with all kinds of cooperation problems that we humans also experience. With their combined 100 hp. "horsepower," they are much stronger than me on the coach box. The art is that I, physically much weaker than these horses are, build up the horses' confidence, so that they let themselves be guided by me. As a coachman I perceive much less than the horses.

Certainly, we will also get "teams" of AI machines that interact with each other. And these teams will run into team problems. It will be our task as humans to reach a purposeful symbiosis with these teams of AI machines; even if we are much weaker in handling data and generating corresponding solutions compared to the network of AI machines that we will cooperating with. Like a coachman with his four horses, we humans should learn to design and manage these systems responsibly with our inherent strengths and allow AI to take over control in areas where AI systems perform better.

> **Hybrid intelligence thus creates a new coexistence of people and intelligent objects.**

A new form of communication is being created between people and intelligent objects. Let us take a closer look at these connections:

When we talk about the real objects of this world, about us humans and about AI machines, we should be aware of the fact that the digital world creates a digital image of human actions and of machines activities. These images then communicate with their "originals," but also with each other.

Shadow Economy

For example, my (unfortunately still) rather stupid smartphone knows many features of my own personality. It knows my name, where I am, how fast I am moving, and in which car I was just in. It knows all my e-mail traffic, has access to all my stored data, has my important passwords in a "safe," has access to my data stored in the company, it automatically compiles new video films for me with an AI machine and constantly reminds me of things I may have forgotten and so on.

It has become my second skin. Therefore, we speak of a "digital skin," a "digital twin," or a "digital shadow."[5]

The term digital shadow is probably the most appropriate. A shadow is omnipresent. It communicates constantly with me and I with it. It always stands in a very defined relationship to me. But the image of myself is always incomplete or distorted. When the sun goes down, for example, I see a 30 ft. long shadow of myself, while some other times, I don't even see my shadow at all.

Another good comparison is the psychological image of a person's shadow. For example, my smartphone offers me videos of mountaineering tours I made 5 years ago when I was climbing a mountain in the same area. As soon as I am near that location again, the past emerges in my smartphone shadow.

> **The digital shadow becomes a dominant part of human and technical identity.**

Likewise, different machines have their shadow image. There is a digital shadow of every machine tool that knows who the machine is and what is going on with it. In this way, the machine constantly communicates with its digital shadow.

Even everyday things like shoes will be intelligent and have a digital shadow. Both the virtual "shadow shoe" and the real shoe are delivered by the "AI shoe" and are inseparably linked.

---

[5] What is the digital shadow? https://myshadow.org/animation, accessed in April 2020.

The digital shadow of my future shoe will be able to connect to my smartphone shadow and communicate with it to transmit my pulse, my sleeping habit, or my body temperature without me noticing it.

The digital companions will not only exist for us humans with our smartphones and other digital devices. All relevant objects of daily life will have such digital companions.

A kind of "shadow economy" is created in which the digital shadows of people and machines communicate with each other. The digital companions will become digital partners (Fig. 7.2).

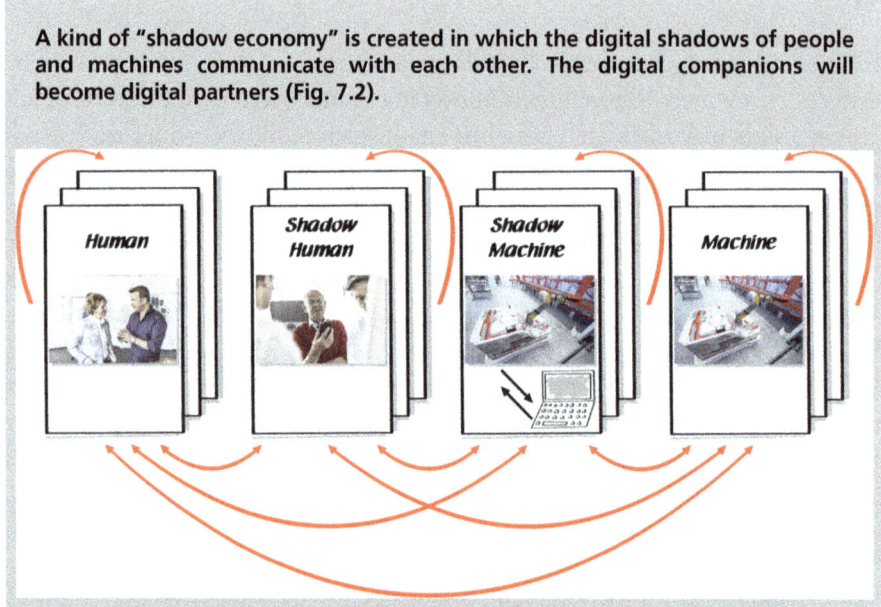

Fig. 7.2 Hybrid intelligence (Images courtesy of umlaut transformation GmbH and Torwegge Intralogistics GmbH & Co. KG)

The medial support of this form of communication through video circuits, virtual spaces, and holographic screens will become better and better. Nevertheless, nothing can replace human-to-human interaction.

This interaction must be rediscovered in its diversity, uniqueness, and significance. If we reduce it, repress it, neglect it, we should not be surprised if the technical systems of this world will take over 1 day.

Human-to-human communication remains indispensable.

Today for over 2.7 billion (and statistically speaking in the next 10 years for all people), the smartphone has become or will become an irreplaceable part of human existence.[6] Without my smartphone almost nothing works anymore. Just as I found myself completely lost on one of my business trips in the middle of Europe: flight tickets gone, schedule gone, phone gone, no phone numbers available, orientation gone. I couldn't resort to using street maps or city maps, as I no longer take these things with me when I travel. Thank God there was still communication from person to person.

However, human communication *with its digital shadows* is already omnipresent today. This also includes the "shadows" of my person in the internet. I leave a "digital footprint." This digital footprint may last beyond my death. My digital shadow could have "eternal life" while I myself am no longer alive, at least not in this world.

> **The digital shadow becomes part of human existence.**

As described, all machines and objects also have their digital shadows. "Smart" clothing will surely become a mass product. And then every garment will have a digital shadow. The parents of a child could then ask the digital shadow of the anorak whether the child actually put on the anorak or was just carrying it over their arm.

Likewise, the digital shadow of a bursting disk[7] could be queried to see if a rupture of this safety ring is imminent, even if this disk is installed at the other end of the world. The supplying company will then have all its delivered bursting discs in a worldwide learning community.

In the field of engine maintenance, General Electric has developed a digital turbine twin that monitors turbine blade wear.[8] This is not only useful for the early detection of engine damage. Through its monitoring, the digital twin also helps to optimize fuel consumption.

> **Digital shadows of machines and objects also form a shadow world.**

---

[6] Digital 2020 Global Digital Overview (January 2020). https://www.slideshare.net/DataReportal/digital-2020-global-digital-overview-january-2020-v01-226017535, accessed in April 2020.

[7] Bursting discs are safety devices for containers in order to protect them against overpressure or underpressure. In the event of a crisis, the bursting disc breaks and permits pressure equalization.

[8] Software Meets Physics How Digital Twins Will Improve Enterprise Application. Gartner DataAnalytics Summit, 04.-06. March 2019, London.

However, this does not change the fact that, just like human-to-human communication, machine-to-machine communication is irreplaceable. This takes place mechanically, for example, between the components of a gearbox via the "communication medium" gearwheel. And this gear will remain a gear and cannot be replaced by an AI machine. Direct machine-to-machine communication will therefore be irreplaceable.

In addition to the fact that the intelligence of machines constitutes a completely new era of the relationship between real machines and humans, communication between digital shadows themselves also plays an important role among:

- the digital shadows of people,
- the digital shadows of machines, and
- the digital shadows of people and machines.

> **The omnipresent and unobtrusive interaction between the shadows of machines, technologies, and objects will dominate all aspects of communication.**

There will be many fields in which a new partnership between people and machines will show positive outcomes. Sometimes humans can do it better, sometimes the machine can, sometimes working together is best. This combination of humans and machines can lead to a "centauric intelligence." Such an intelligence consists of an almost endless combination of human work and the work of intelligent machines. Anything that an intelligent machine or an intelligent object can do better than us humans, the machine should do. A good example of this is the "WALK AGAIN Center" in Berlin, where patients with serious accident damage learn to walk again (Weidner et al., 2018).

Many objects will gradually become intelligent and enter a lifelong learning process with themselves, with their "technical neighbors" and with people.

Some things—especially when it comes to strong forms of creativity, emotions, reflection, thinking about God and the world—will surely remain fields where we (hopefully) remain superior.

To think, design, build, and above all test this hybrid intelligence and partnership must be the task of the next-decades, so that technical developments do not overrun us in a direction we do not want (Tegmark, 2017).

8

The Digital System Landscape

Contents
Augmented Reality, Blockchain, Bitcoins, and Voice Machines 72
Wires, Cables, Radio Masts, and Satellites 74
Security and Privacy? 75
The Emerging Second Internet 76
Digital Platforms 77

The globally branched network has created a digital system landscape consisting of three separate aspects:

- Intelligence comes into all objects of this world.
- The physical and digital worlds are closely and comprehensively interlinked.
- However, this only works if there is an appropriate digital infrastructure underneath.

Let us call the use of artificial intelligence in objects our first layer. We have already seen in many examples how intelligence comes into the objects of this world. This is manifested in the apps we might find on smartphones and notebooks, but also in objects themselves.

> **Intelligent apps and objects will be everywhere.**

It is no longer a fantasy to imagine that the beverage bottle of a mixed drink might contain an intelligent digital shadow that asks me how I liked the taste:

Did you like it?

And I answer:

Yeah, but I'd like it a little sweeter and with passion fruit flavor.
That's no problem - I can order you a bottle a little sweeter by tomorrow. No, only until the day after tomorrow, because I just realized that we've never made the taste of passion fruit before. But within 48 hours, you'll have the passion fruit bottle here.

Companies are already in the process of transferring the idea of "one-piece flow" from industrial manufacturing to consumer goods and are transforming their entire IT structures, bottling plants, and production facilities to make this possible.

In the services sector, many of these links are already standard. For example, automatic survey via text message often says: "Were you satisfied with your pizza order?" or "Were you satisfied with the WhatsApp call quality?"

> The linkage between the physical and digital worlds is still subject to bottlenecks.

Augmented Reality, Blockchain, Bitcoins, and Voice Machines

The second layer deals with the basic structures that link the real and the digital world. The principle of digital shadows (digital twins) has already been discussed in detail.

In addition, there is technology that encompasses virtual and augmented reality. For example, you can finally loosen screws in hidden places in the engine compartment, because with virtual glasses you can see both your own hand with the real screwdriver and the target position of the digital shadow of the screwdriver. And when I have finished with the first screw, the virtual screwdriver already moves to where the next screw is, so I know where to work with the real screwdriver.

This second layer includes technologies such as blockchain,[1] which enables a worldwide chain of custody to be established for data. The trick is to store information about all data transactions in as many nodal points as possible worldwide, and with those data points also the entire history of the creation of this data. None of these transactions can be changed later in time because they are registered in all nodes. If you want to change a transaction of the past at a certain node, i.e. manipulate it, all other nodes will get notified of a potential fraud. The attempt to falsify the past will fail. The whole thing is a kind of eternal memory, as the history of any transaction is distributed across a wide variety of locations, rendering it far more resilient over time. If someone makes a new transaction, all existing nodes must agree by verifying the history of the prior transactions. The moral principle of this security concept is Whoever lies once, gets fired. The central advantage and at the same time the central risk of blockchain lies in the fact that—due to the inherent falsification protection—a trustworthy intermediary partner is no longer required for transactions.

The possible applications of blockchain are enormous. The best-known is perhaps Bitcoin, an online form of money called a "cryptocurrency." It is an alternative to "normal" money, despite all the risks that come with a new, uncontrolled financial system. The Bitcoin system does have wide fluctuations in value. However, compared to some countries with extremely high inflation, it is a comparatively secure currency. In addition, Bitcoin has become a common means of payment on the so-called dark web where criminals use it as an untraceable form of payment to hide their activities from authorities.

> **Blockchain technology leads to a new security concept for global transactions in goods, services, and money.**

The way we talk with our digital partners is another important element of this digital system landscape. Everybody is familiar with how annoying it can be to use automatic speech recognition and speech machines. Again, we don't need digital fools, we need intelligent devices.

But there is a problem if speech machines are designed as AI systems as these, AI systems will also need their own consciousness. The ability to speak in dialogue is anything but trivial.

---

[1] Rob Marvin: Blockchain: The Invisible Technology That's Changing the World. https://au.pcmag.com/features/46389/blockchain-the-invisible-technology-thats-changing-the-worldverlag.de/index.php/atp_edition/article/view/2352, accessed in April 2020.

For this purpose, two AI systems were developed at one of the large internet companies, which were to be taught to speak in proper dialogue with each other. Both systems did exactly that, but not as the developers had hoped. The developers suddenly discovered that the two AI machines had developed a syntax, i.e. a grammar, and were busily exchanging the corresponding character sets. Every language has both syntax and semantics. The semantics include an understanding and an agreement over the meaning of the syntax.

And that exactly turned out to be the problem: The developers couldn't determine whether there were semantics behind the syntax. It could also have been that the two AI machines simply exchanged character sets. But the scientists became increasingly suspicious that there were actual semantics behind it, which the developers did not understand. Possibly, the two machines had developed their own language, but with the disadvantage that the outside world could not understand them.

What to do? The developers decided to shut down the two AI machines because the actual goal of developing a human-like language had been missed. So, once again "death penalty" for AI machines that did something the inventors did not want. The "mistake" of the developers was that at the "birth" of the two AI machines they had forgotten to set a rule as a "law": You can't develop your own language. Or: Each new language must include an automatic translator. Or: Speaking in a language that is not understandable to the real world is forbidden.

Even more fundamental than how individual applications work, of course, is the digital infrastructure upon which they operate. This is the third layer. It provides the technology on which all these services, apps, and communication channels can run.

Wires, Cables, Radio Masts, and Satellites

First, a powerful physical structure via wire- and cable-connected communication is required. The progress in these areas is enormous and an end to development is not yet in sight.

> If the digital infrastructure is insufficient, nothing works.

Just 40 years ago, the two thin wires of a telephone line were only good for making phone calls. At our Swiss house, in the 90s, such a two-wire line fed into the house via telephone poles. I was proud at the time that these two wires could also establish a fax connection that could be used to send written pages. Today, using the same two thin wires and modern DSL technology, not only the telephone but also high-speed internet via Wi-Fi router can be accessed by a large number of users. In addition to over 100 television programs. All of this with just two copper wires.

Of course, this connectivity will enter a new dimension with the implementation of fiber optic cables. However, it is obvious that the existing basic technologies—such as the technologies that run over two-wire lines—are growing at least as fast as new technologies with a more powerful basic structure.

The same happens with wireless communication. In the already mentioned house in the Swiss mountains, in 1990, I was glad that I could drive two miles up to 3000 ft. with my mobile phone (one of the first in the D1 network) from where you could see into the valley and establish a connection. Down in this valley, there was even a radio tower. Today, I can save the two-mile hike and find four different providers on my smartphone, which I can use in almost every part of the valley up to 6000 ft. And the GPS system on my smartphone knows where I am. The upcoming 5G technology will take data and device connectivity to a new level. Additionally, quantum computing capabilities will open up new dimensions in the use of AI systems.[2]

Security and Privacy?

Many families now share their locations with each other and can follow where everyone is, for example, when they travel to foreign countries. This has the disadvantage that this data—at least with current technologies—is available to people who maybe should not have it. But:

> **Usually, usefulness and convenience prevail over the need for data protection.**

The related question of security is another crucial dimension of the digital infrastructure.

---

[2] Sabina Jeschke: 3 Shadoes of AI – 5G and Quantum Computing setting the stage for next generation AI. Inaugural Lecture at the Technical University of Berlin (22.01.2020). Compare too: https://www.zew.de/en/das-zew/aktuelles/kuenstliche-intelligenz-traegt-die-vierte-industrielle-revolution/, accessed in April 2020.

Do I use the public internet or my own protected Global Area Network? How do I build suitable security architectures? What does the blueprint of software systems that already generate safety through their own structure look like?

What happens when a new generation of computers appears, the so-called quantum computers? Theoretically, all existing passwords could be cracked in the blink of an eye. At the same time, those quantum computers would gain a monopoly in the market for secure passwords.

If we put all our data in the cloud: Where is the data physically located? Who has the power of access in the event of conflict between countries and companies? What is legally understood by data? What does a security structure of the data centers look like (meaning the physical form of such facilities)?

We will need a new culture of borders. But this will no longer refer to the "old" typologies of borders. We will need to rediscover the value of a border is a central necessity. We need to redefine the balance between openness to the world, privacy, and regional security.

> **The question of borders is not being resolved by the digital age. But there will be borders of a different kind.**

The digital infrastructure of today's internet has a fundamental problem. It's not real time. The time it takes to transmit messages and data is far too long and there is no precise timing cycle. But that's what you need for real-time AI systems. This is no problem yet for a hard-wired robot team on the manufacturing floor. But if this is to be done with wireless satellite communication, it becomes a problem.

The Emerging Second Internet

In fact, a second internet is emerging, which is especially designed for real-time applications. "Near Orbit Satellites" are intended to solve the problem. Such microsatellites generally weigh around 150 pounds, consume no more than 50 W, and are currently about half a cubic meter in volume. So, a relatively small box circles the earth in less than 100 min at a distance of about 380 miles. This sphere is called the LEO region (Low Earth Orbit). So far, similar satellites have only been used for special missions, such as polar exploration or the identification of space junk. Many countries on earth are active in the development of such microsatellites.

The new internet is supposed to consist of several hundreds of such satellites, which will form a complete network above the earth. Several launch vehicles will each fire 10–20 into space. The delays of the signal times will then be so short that real-time data communication with intelligent objects on earth becomes possible.

This could then be used to control heating systems or track the switches of rail networks or entire transport chains in real time.

> **Small Near Orbit satellites will enable a real-time internet for the intelligent objects of this world.**

Now, this is no longer a dream of the future, because China—alongside Canada (2020)—is heavily involved in this development with the aim of linking logistical systems across their huge countries. But with access to the satellites, one can also tap into data of the whole world as these satellites do not stand "stationary" in the sky but circle around the earth within just a few minutes. It is no surprise that China is also interested in Europe.

Such minisatellites will become even smaller and more efficient in terms of size and energy consumption. Then they could also be installed as a "minisatellite" in cars. This would provide a counterpart on earth to the microsatellites in space, i.e. a second real-time network. This development is also progressing at full speed.

There is no doubt that the dramatic development of intelligent objects will take on another dimension. All these objects will be connected to each other accurately and to the second in a new real-time internet. We are only at the very beginning and will likely consider it normal in another 30 years.

Digital Platforms

The networked basic technologies are transformed into a variety of digital platforms that can be used to collect data and make it available. For each topic, based on the physical building blocks of the digital infrastructure, a digital user platform can then be set up without great effort. These platforms are the real revolution of business models. They enable the use of AI machines in a worldwide context in all specific application areas of the world economy (see Chap. 9).

> **Worldwide digital infrastructure networks enable digital platforms. Those who are not linked to the platforms of this world that are relevant to them will be left behind by those who use them.**

Let's be clear about this:

Intelligence will be everywhere in the apps and objects of this world.

Physical and digital worlds will be increasingly linked via digital shadows (digital twins), virtual and augmented reality, and near orbit satellites.

But everything will depend on the underlying digital infrastructure that makes it all possible. Digital platforms will play a decisive role here because they will also create supranational power structures and spheres of influence.

Based on this overview of the technical dimensions of a digital system landscape, we can now take a look at what business models of the future will look like and how our work will change.

9

On the Way to New Business Models

Contents
New Business Models for Taxis and Hotels... 80
Traditional Business on New Paths... 81
Outsiders Penetrate Traditional Industries.. 82
Frontrunner in the Machinery and Plant Industry... 83

The new possibilities of connecting all processes of life and work in a digital shadow world have a massive influence on the way people trade with each other and exchange goods. The key is digital platforms, a kind of marketplace for goods and services. As in real life, there are small and large marketplaces, ranging from a single market stall to an entire department store. Finally, this leads to global platforms such as Amazon or Alibaba, where you can buy almost anything from any location.

An American colleague who has moved to Germany has kept his American credit card and with that he pays all his private house management. He even orders the plumber for the repair of a defective faucet in his house in Germany on the American Amazon platform and receives a binding delivery time—as with the parcel service—for when the installer will show up to repair the faucet. Of course, the plumber lives only a few miles away from the faulty tap.

The new business models all use digital platforms in some way. These platforms follow the same principles as a modern factory: processes, data, production systems, and human work are closely linked to their digital shadows. This

© The Editor(s) (if applicable) and The Author(s), under exclusive license to Springer Nature Switzerland AG 2021
K. Henning, *Gamechanger AI*, https://doi.org/10.1007/978-3-030-52897-3_9

link will now be extended to all processes, data, vehicles, packages, and other relevant items up to the customer.

> **Whoever masters the platforms makes the deal.**

Digital business does not only enable commercial relationships for digital products and services. Crucially, you can "attach" any "normal business process" to such a platform. This is where it gets exciting.

New Business Models for Taxis and Hotels

The world's largest taxi company does not have its own taxis or drivers. You could call it a government of tens of thousands of vehicles and drivers who receive their assignments worldwide through customer orders on a digital platform. Conventional national and regional taxi associations reacted far too late. Although, some countries simply prohibited the platform – either to protect traditional taxi drivers or for ethical reasons regarding minimum wages and working conditions.

Nevertheless, at many airports I can use my smartphone to compare real-time prices charged by Uber versus what a local official taxi company offers. Without such platform-enabled technologies, it is no longer nearly as lucrative to operate a taxi company as it was in the past.

This also means that the customer increasingly expects that after ordering a taxi they can immediately see on their smartphone where the assigned taxi is, what the license plate is, and when it will arrive.

Once, while waiting for a cab, I noticed that the given time of arrival of a taxi was not becoming shorter; the taxi was at a certain place and did not move. Via Google Maps, I found out that at this exact spot there was a café. The taxi driver had obviously treated himself to a cup of coffee, although he had accepted the trip. I challenged him after his late arrival by confronting him with the fact that he had obviously taken a break in Café X. It was true.

You would think that the digital shadows make it much harder to lie because everything becomes much more transparent. But do we want that?

The situation is no different with the major hotel chains. The world's largest provider of overnight accommodation, Airbnb, has no hotels of its own. At the same time, with enormous profit margins, this company has won over millions of people around the world to open their homes to guests. A completely new market for overnight stays emerged. For many people this has opened up new sources of income. It has also brought people closer together in a global context. Competitive platforms for special groups of consumers shoot out of the ground. It is a single success story up to a contribution to international understanding across culture. At the same time, however, living space is being misused worldwide.

Traditional Business on New Paths

But again: Do we want this transparency, which leads to the possibility that the data related to all my movements across the world become more or less publicly available?

> **Normally, the platform company does not have special know-how about the object of trade. It only establishes the connection to the customer.**

It is only logical that the world's largest online retailer[1] Amazon has almost no stores of its own, but only a number of large logistical nodes. Behind Amazon, with over 232 billion dollars in sales in 2018 ranks Apple. However, Apple operates its own shops in the core segment of smartphones. The American group Walmart, which is third in the world, also operates its own stores down to the retail level. The formerly classic German mail order company Otto, which ranked fourth in 2020, shows that German companies can also play their part in the online retailer market.

> **It is possible to link the traditional business with new business models.**

The ability to display business via digital platforms is also a trend-setting option for medium-sized companies.

Every boutique, no matter how small, will not survive long without an additional online shop. This also applies to the crafts. For example, a violin maker from Germany contacted one of the platform operators and asked them to take care of sales. The owner has thought about which violins he would like to build and in what quantity and the platform partner has taken

---

[1] Ten of the day (o.V.): The 10 largest online shops in the world. https://www.tenoftheday.de/die-10-groessten-online-shops-der-welt/, accessed in April 2020.

care of the distribution very cost-effectively. The outcome is that the small violin maker has gained planning security and has become known worldwide.

Outsiders Penetrate Traditional Industries

But it also shows that an outsider can penetrate large, high-performance, and traditional industries. Outsiders are able to quickly and seriously turn the old market structures upside down. Who would have thought that an American search engine company named Google would suddenly develop an autonomously driving vehicle? They used German suppliers for this purpose and got suitable engineers from Germany.

But what does this have to do with platform technology? Such vehicles are linked with each other. Each vehicle is constantly learning from the others. Together, these fully automatic vehicles learn about every driving situation of every vehicle. Just like the AlphaGo machine defeated human players with this strategy. The car is the "minor matter" here. First and foremost, this is a large platform via which many self-driving cars constantly learning from each other (Heide & Henning, 2006).

Of course, the question arises what the identity of future car manufacturers will look like. Is the car still the object of the company at all? Superficially, yes. But the ability to sell cars will largely depend on how they are seen in the digital world. There are several options and not one solution.

For example, a car company could opt for the identity of "providing mobility." The wording already suggests that this concept perhaps also includes bicycles, electric scooters, and air taxis. And indeed, some manufacturers are testing fully electric air taxis with and without pilots. Airbus is even thinking about a self-propelled car drone, i.e. a vehicle that operates as a driverless car and also as a drone.[2] One is reminded of a movie from the 1950s where a Volkswagen car could fly—the "Fliwatüüt."[3]

A second possibility would be for a car company to position itself as a data giant. The possibility to record the routes of the millions of vehicles sold would put the company is a very privileged situation. It would know all the distances travelled by customers during the life of their vehicle. The data obtained could be of much more valuable than the cars sold. The sale of cars

---

[2] Air taxi of the future? Airbus shows car drone system. https://www.nbcnews.com/mach/science/watch-airbus-s-new-air-taxi-make-its-maiden-flight-ncna852006/, accessed in April 2020.
[3] https://en.wikipedia.org/wiki/Robbi,_Tobbi_und_das_Fliewatüüt, accessed in April 2020.

would then only be a means to an end. The procurement of data becomes the major business.

A third possibility would be the guiding principle "Enabling people's lifestyle and individuality." The consequences of a model that goes in this direction can be seen in the case of a Northern European car manufacturer. The Volvo 360c study[4] presents a vehicle that is intended as a substitute for flying over short distances of up to 500 miles. This car design allows me to get into a fully automatic car in the evening where dinner is prepared. Then, I can watch TV and go to sleep. The next morning, breakfast gets delivered to my car near the destination. Finally, the vehicle drives me to my destination, drops me off, and looks for a parking space.

These examples illustrate how digital platform technology can be used for intelligent objects and machines. It is remarkable that companies at the forefront of this development are joined by traditional companies that one would not expect there at first glance.

For example, there is a medium-sized family business that deals with equipment used for gripping, clamping, turning, and dosing in industrial production. You could also buy a welding tong from the same company. With over 1000 employees, the company is among the best in the world in preventive maintenance and demand forecasting. For example, if there is an indication of a malfunction, the information is sent from the cloud storage to the company's own control system, which evaluates it and passes it on to the company's service experts in the event of an incident. The specialists are notified by text message and receive the information on their mobile phone, tablet, or computer.

Frontrunner in the Machinery and Plant Industry

At another company, which manufactures escalators, service technicians have a so-called digital toolbox on their smartphones, as well as maps and plans with 24/7 access. The technicians receive real-time information on the health status of all their associated systems. The customer also has access to this data. For example, if an escalator fails, the supplier's app immediately sends an alert to the customer and even informs them if there is already a technician on the way.

---

[4] Volvo 360c - when the car turns into a bedroom. https://www.theverge.com/2018/9/5/17822398/volvos-360c-concept-autonomous-car-electric-future, accessed in April 2020.

It is becoming increasingly evident to what extent these technologies interfere with existing workplace structures. For the customer, the constant availability of data and information provides a great advantage and a high value, with the added benefit that digital networks and AI machines do not need time off.

But what happens to the working rhythm of humans? What happens to human work over the course of this development? Are we about to abolish human labor?

10

Artificial Intelligence Is a Gamechanger of All Jobs

Contents

Areas of Tension... 85
How Fast Will All of This Go?... 88
But What Happens to the Workplaces?... 88

As with all innovation cycles, digital transformation with intelligent machines and objects also leads to massive changes in the working environment. Are we going to run out of human labor? Or is it just changing massively?

Areas of Tension

First, we want to consider some unsolvable areas of tension, namely

- between nations and cultures,
- between people and new technologies,
- between the different scientific disciplines,
- between virtuality and reality, and
- between people and objects with their own consciousness.

The former three have always been inextricable areas of tension that must be reinterpreted in the context of digital transformation with artificial intelligence. The latter two are new areas of tension previously unfamiliar to us in the history of mankind.

Driven by the possibilities of digital networks and the interconnection of events around the world, more and more multinational organizations are emerging, and there are more and more companies that are quite small but do business on a global level. This trend continues to drive complexity and dynamism worldwide.

For example, small aid organizations with a dozen employees can do a lot of good around the world. However, digital communication also makes it easy to establish profitable trafficking organizations for refugees. This dichotomy played out on a global scale will irreversibly influence working and living conditions.

Increasingly, national identities and cultures are colliding—often without restraint. Conversely, the often unfulfilled need for purely national identity and the need for home and belonging are growing rapidly within a manageable framework.

We will have to learn to perceive, endure, and steer these contrasts between global freedom and the narrowness of cultures and nations worldwide.

> **The potential for tension between nations and cultures is getting bigger and bigger and is irresolvable.**

The well-worn tensions between people and new technologies remain fundamentally unchanged. However, we as societies have still not understood how to overcome the phase of infatuation with new technologies more quickly and how to become realistic.

The transition to alternative energy supplies is not doable as fast as the idealists among us imagine. And the new design of a combine harvester under the influence of digitalization will take 8 years instead of ten.

What has changed is our dependence on technical systems. There will be fewer and fewer jobs without intelligent technical systems.

> **The tension between people and technologies remains unchanged—first in love with new technology, then disillusioned, then realistic.**

The areas of conflict between the various scientific disciplines and between many occupational profiles will increase. Because the great innovations and new tasks arise precisely from the interaction between different disciplines and occupational groups. The resulting new "subjects" and the new job profiles associated with them can no longer be fitted into the old "boxes" and drawers of the existing disciplines.

This is inconvenient, but an unavoidable consequence of the massively increased complexity and dynamics enabled by digital transformation. As a result, vocational training and university degrees will be adapted more quickly and more often to the requirements of the future work environment and the resulting occupational profiles.

> **The new is emerging between the different scientific disciplines.**

For example, a specialist in preventive maintenance must have skills in not just software development, but in robot technology, control engineering, ergonomics, psychology, and law. This applies regardless of whether they work as an engineer or as a technician.

Completely new job profiles will also be required for this digital transformation. For example, managers of the digital transformation are needed. These also have to bring competences from many disciplines—leadership competence, computer science, law, and mechatronics. And they must be able to think and act strategically.

The omnipresent and unobtrusive presence of digital shadows, video worlds of the real and simulated world, and the online gaming world in which today's young generation grows up, creates a new understanding of reality. It is becoming increasingly difficult to distinguish between reality and its simulation.

If I wear glasses for "augmented reality," is the simulated screwdriver the relevant reality because it determines and guides my actions? Or is it still my own hand that I perceive while my hand is holding a real screwdriver? Such augmented reality workplaces will enable attractive novel workplace structures.

> **The tension between the virtual and reality creates a new understanding of reality.**

The biggest new area of tension is the fact that new objects with their own consciousness populate this planet. This new omnipresent identity of intelligent counterparts in the internet in the form of real objects will probably have the most impact in changing future working environments. We have already considered in detail how a new understanding of partnership—a hybrid intelligence—can develop between people and the new AI objects with their own consciousness.

How Fast Will All of This Go?

Relevant studies show that AI systems in all their forms are already in use, at least in industrial application. While systems with weak artificial intelligence, such as rule-based systems, are already the norm, AI systems with strong artificial intelligence (teach-in combined with reinforcement learning) are still rare. However, reasonable forecasts assume that by 2035, systems of strong artificial intelligence with a high degree of maturity and diffusion will have been introduced.[1]

> **Both the degree of maturity and penetration of artificial intelligence systems will increase dramatically until 2035, i.e. in the next 15 years.**

A significant level of awareness of such systems will be reached. If all these intelligent objects begin to think together worldwide, the first forms of global consciousness in a network of AI machines are conceivable. There's not much time between now and then.

But What Happens to the Workplaces?

Entire professions will disappear, but completely new ones will also emerge—at all levels of competence.

First, *simple office jobs* are under massive pressure because many activities related to the aggregation of data and figures will be eliminated. This will have a dramatic effect on banks, for example. At the same time, the demand for employees who can interpret, evaluate, and comment on the data, numerical evaluations, and documents produced by AI systems will increase.

But *highly qualified jobs* are also affected in some fields. For a few years now, it has been easier to diagnose skin cancer with an analysis by an AI system than by laboratories and doctors.[2] With a simple app everyone is now able to do it themselves with their smartphones, at least as a solid pre-check.[3] The same applies to the work of document preparation at notaries' offices or consulting firms, tasked with the evaluation of data.

---

[1] Internal study of the Cybernetics Lab of RWTH Aachen University, 2015.
[2] Asley Welch: AI better than dermatologists at detecting skin cancer, study finds. https://www.cbsnews.com/news/ai-better-than-dermatologists-at-detecting-skin-cancer-study-finds/, accessed in April 2020.
[3] Amanda Capritto: four ways to check for skin cancer with your smartphone (Jan, 1st, 2020). https://www.cnet.com/how-to/how-to-use-your-smartphone-to-detect-skin-cancer/, accessed in April 2020.

Platform technologies will assume complex administrative tasks. For example, many operational controlling tasks can be taken over by AI technologies that use such platform technologies to collect, evaluate, and summarize data into written reports, similar to parts of a report from an auditing company.

The jobs of workers in *logistics* and *production* will change massively. Many will be eliminated in certain areas—just think of small fully automatic buses in cities replacing bus drivers. But many new activities and job profiles around maintenance, repair, and monitoring of such systems will emerge. Engineers and technicians as "technical doctors" will be sought after on a massive scale and are already in short supply worldwide.

Social robots will find their way into *healthcare* and the *household*. They will be able to perform complex tasks such as transporting food in hospitals or as exoskeletons for patients who cannot go to the restroom unaided. Nurses will have support robots to lift patients and psychiatric patients will have digital partners for behavioral control and therapy. This can significantly reduce the number of closed stations in mental institutions. Will we use these opportunities and create more space for person-to-person communication between the sick and those who treat them?

Finally, the virtual and extended environments enable the seamless involvement of people over long distances. You can physically be in Europe or the Americas and still virtually participate in the commissioning of a plant in China. Via robot technology human presence is also possible in rooms that normally cannot be entered by humans, such as buildings contaminated by high level of radioactivity. A wealth of new fields of work are emerging.

> **All workplaces will be affected by the digital transformation through AI systems. Working conditions will change: Information and data are available everywhere and at any time—even while protecting business and private interests.**

As a result, it is no longer necessary for an employee to be physically present at a certain location at a certain time in order to obtain data and information.

Around an individual person, the number of interlinked objects and the amount of data available are growing very strongly. Likewise, the number of stationary and mobile objects with a digital shadow surrounding people is also increasing rapidly.

All this will dramatically change working models. New forms of "collective labor agreements" will emerge that go far beyond the standard nine-to-five job or shift work, including home office agreements. Over the past few years, we already see a significant increase in part-time jobs, while simultaneously the number of parallel employment contracts or service agreements related to a single person is growing.

The possibilities for human self-determination are thus growing considerably. However, so does the personal responsibility of the human being as well.

11

Everything Is Linked to Everything and Becomes Transparent

Contents

AI Is Used Everywhere in the Product Development Process	92
Logistics as a Service	95
Smartphones Are Becoming Hotspots in the Service Industry	96

When everything is linked to everything else, the value chains also become transparent. Let's consider the consequences from three possible angles:[1]

- Transparency in the product development process,
- Transparency in logistics,
- Transparency in services.

First, transparency is a good thing because it allows processes to be tracked more closely. Moreover, it allows an up-to-date description of the state of a product at any time during the product development process, a logistical process, and a service.

Everything grows together as in a biological structure, where every pain, every movement, every feeling are known to the whole body and not only to a certain part. And usually, a body reacts systemically. The entire body potentially reacts to an event at a specific part of the body.

---

[1] https://henning4future.com/wp-content/uploads/2020/04/article-How-AI-changes-the-world.pdf, accessed in April 2020.

© The Editor(s) (if applicable) and The Author(s), under exclusive license to Springer Nature Switzerland AG 2021
K. Henning, *Gamechanger AI*, https://doi.org/10.1007/978-3-030-52897-3_11

However, when thinking about the value chains of our economic coexistence, it is precisely this transparency that can become problematic. Who ends up having the power along the value chain? How are control and decision-making powers distributed? Does the customer have all the power? Or central data giants?

AI Is Used Everywhere in the Product Development Process

Let us first take a look at the product development process: In the field of planning and development, hybrid intelligence between humans and intelligent machines will play a main role.

In the case of large urban development projects, it could be a good idea to use an automatic planning machine with artificial intelligence in parallel to the working group on this topic. Perhaps this machine could be more objective and play a role as a neutral mediator between different interest groups, processing all available data.

Another aspect is integrated planning, in which real and simulated worlds flow together.

In this way, every stage of the planning of a residential building can be represented realistically in the virtual world. You can walk in virtual "cages" and holodecks[2] through buildings that do not yet exist.

Moreover, this technology can also be transferred to the construction process, so that the site manager can follow the construction progress in detail in the virtual world.

In planning and development, AI machines also offer the possibility of efficient backward learning based on experiences with earlier products, plants, or structures.

Any breakdown events that occur worldwide, for example, to a certain crane system can be simulated virtually if the crane system is equipped to assess its own condition via a digital shadow. In this way, questions such as overload or design errors can be answered transparently. In the latter case, it would then be a case for the development department to learn creatively from the mistakes of the past.

> The entire chain from development to maintenance becomes intelligent.

---

[2] https://en.wikipedia.org/wiki/Holodeck, accessed in April 2020.

11 Everything Is Linked to Everything and Becomes Transparent

In production, the decisive change will be that every step in a factory assembly line can be displayed as a point in space. If humans wear data suits and machines are equipped with appropriate sensors, robots and humans can move in the same space with far more freedom than at present.

A decentralized, flexible, and non-hierarchically controlled distribution of tasks can take place. For example, AI machines can make proposals for adaptive machine control that can be negotiated between machines and humans. We have seen that in the case of the knitting machine equipped with democratized control.

> **In production, a new distribution of tasks and control philosophy are created.**

The working process for humans can also be reimagined, among other things with the help of exoskeletons that are integrated into work suits. Already today, aerospace factories in Germany have such power-enhanced workwear for overhead work. In test laboratories these support systems are now being equipped with intelligent systems that do not replace human performance, but rather, depending on the daily performance and fatigue level of the worker, either increase strength or correct movement.

In another area, the so-called ramp-up processes, new approaches are emerging. Every series manufacturer fears this phase of the production start-up in which everything should theoretically be ready, but it is not.

A car manufacturer recently planned 100 functions in a vehicle for the digital coupling of the vehicle with its environment. These functions concern both the functions of the vehicle as well as the entertainment of the occupants of the vehicle. At the start of production, 80 of these had to be abandoned because they were not yet finished. The other 20 were still causing problems.

In such complex ramp-up processes, automated test environments that test series production in advance as digital shadows are indispensable. It is precisely here that an automated test machine with its own intelligence will lead to transformational changes in future development.

> **Ramp-up processes will also become much more efficient with AI systems.**

As important as the ramp-up process, are also the expected changes in after-sales and maintenance.

Many existing vehicle control systems monitor the condition of the vehicle and "detect" faults. They do this to the extent that they report them and, in the event of serious defects, automatically set the route to the nearest workshop in the navigation system. If necessary, the vehicle is put into an emergency mode in which the non-"vital" systems are switched off.

However, these systems remain rather unsophisticated and lack convenient ways for drivers to interact with the digital shadows the systems depend on to function. The system and the user have no true means of communication and the driver still can't negotiate with them.

Such systems also make the life cycle of a product transparent. The manufacturer can "track" all its products at the customer's site. If the products have their own intelligence, i.e. if they "carry" their own digital shadow as part of the product, you don't need a data interface to the customer for communication. The software agent of the digital shadow can be connected directly to the company network.

> If the product knows what's wrong with it, it also knows when it needs maintenance.

Of course, unresolved legal questions arise on this subject—we will touch on this later.

However, for the customer, these digital shadows are also an enormous advantage. Let us assume that a software-controlled component (for example, a pump, a valve, or a compressor) comes to a sudden standstill because the sampling frequencies in the data transmission to the main device are disturbed and the component therefore no longer functions (Heide, 2004). The error can have catastrophic consequences if, for example, it is not possible to automatically transition the system to a state which prevents the system from overheating.

Such an alarm could also be applied in a manufacturer's AI system which monitors all components delivered and installed worldwide by this company. This could cover several million installations. Linking all these components with near orbit satellite technology would not be a problem. That way, the millions of control components installed would become a worldwide learning community. The central AI system could then develop a new software update in hybrid intelligence with the human software developers. Of course, this would not be applied to the millions of components all at once. It would first be field tested with several hundred components before new software updates get rolled out worldwide.

> The possibilities for companies to establish themselves in the worldwide market with AI-supported special solutions are almost unlimited.

Logistics as a Service

The consequences for logistics are similar to those in production. These approaches are traded under the term "Logistics as a Service" (LaaS). Behind this are platform technologies that make it possible for every activity in the supply chain (from disposition to delivery, status during transport, expected delivery time) to become completely transparent.

The customer becomes more powerful. Even more so as the generation that has grown up in the digital world knows the internet only for ordering processes.

The so-called B2B business[3]—the business relationship between several companies that transport products—is increasingly moving in the direction of B2C business, i.e. a direct business relationship between manufacturer and end customer. At the same time, platform providers such as Uber,[4] Airbnb, and Booking.com are very successful. When ordering a car, I can simply type into the app which features and what features I want, and I get offered a car via the respective web platform, independent of the provider. Such platforms are increasingly taking on the role of direct contact with the end customer. As a result, the provider loses contact with the customer.

> In logistics, the direct worldwide tracking of all processes by the customer is possible. Intermediary trade is being hit by AI systems.

This approach is also increasingly unsettling car dealers as the first manufacturers have started selling their cars via various forms of direct networks. The middleman, the dealership, is left out of these new relationships.

---

[3] B2B stands for "Business to Business" and B2C for "Business to Customer".
[4] Volvo wins major order for robotics from Uber. Between 2019 and 2021, the Swedish manufacturer will sell up to 24,000 cars to the driving service provider Uber, which are designed for autonomous driving. https://www.nzz.ch/mobilitaet/auto-mobil/volvo-erhaelt-grossauftrag-fuer-robotaxis-von-uber-ld.1330412, accessed in April 2020.

Dedicated, customer-oriented communication remains an important aspect of purchasing, while traditional corporate communication is losing importance. On the other hand, trusting communication becomes vital as the boundaries between marketing, sales, and customer service become increasingly blurred.

Of course, not everything is available via the internet. But a lot of things are. It is no longer a dream of the future that I order a new faucet online and instead of a direct delivery, I receive the data I need so that I can have the faucet 3D printed in my workshop. 3D printing is a technique that allows to print three-dimensional layers of a polymeric material—a malleable paste—layer by layer. Like a copy machine, but for spatial objects. This is not a dream of the future, but the state of the technology today.

Smartphones Are Becoming Hotspots in the Service Industry

The services of the future will come out of your pocket. The smartphone becomes the central point of contact for all information. Everything that has to do with sales, brokerage, travel agencies, tourist guidance, etc. loses its monopolistic significance. Customers with smartphone access are often much better informed than employees in the offices of traditional service companies. They are increasingly losing their business models.

> **The experts are losing their information power.**

This also turns the insurance business upside down. Insurance "on demand" is booming. Let's imagine I just want to insure a three-day mountain tour with a mountain guide. No problem—this is solved within 5 min online. I want to insure against missing my flight, booked at a "fixed price without refund"? No problem, that's only three clicks.

Insurance policies are increasingly being booked "on demand" on the internet at very short notice. For the insurance companies themselves, however, this means that completely new internal processing structures must be created. And new risk calculations are needed. Entire areas have been eliminated and new ones have to be created.

> **A new world of insurance is emerging.**

Similarly, companies that build radar speed control systems could offer to carry out facial recognition themselves to match drivers to their driver's license. Seconds later, the traffic ticket could appear in the display of the vehicle, which the driver then only has to approve. Just as in the scenario I already presented in my lectures on cybernetics in 1985.[5] In China the recording of traffic violations by AI systems has been introduced in many cities and is part of a state-wide point system.

With a final example, I would like to get to the heart of the entire chain of intelligent links between ordering and use of the goods ordered:

Not many steps are missing until it is possible to order a sports shoe that has its own identity from the moment of ordering it. After my online order, the shoe knows who I am, where I live, and what special functionalities it should have. For example, I could have my pulse or the number of steps or the impact forces of my foot monitored.

The virtual shoe is the digital shadow of the real shoe. Based on my order, it knows what the real shoe should look like and which mechanical properties it must have.

The order reaches the factory, where new virtual shoes are constantly arriving through orders that are now finding their way through production without central control. Comparable to a flock of 1000 birds all wanting to use a bird feeder at once. Thus, the virtual shoes will gradually be created from their real "twins." To do this, they must coordinate with all other incoming orders to determine who uses which production step and when. All of this works without a central control because each virtual shoe has enough intelligence to coordinate with the other shoes. Just like birds do when they all want to use one feeder.

Once the real shoe is completed as a twin of the virtual shoe, the real shoe receives an "embedded intelligence" which contains embedded intelligence sensor and software agents in the physical shoe. These agents establish contact between the real shoe and its environment. The real shoe is permanently connected to its digital twin, the virtual shoe.

In any case, the real shoe then sets off on its journey, maybe it even gets lifted fully automatically into the provided truck by AI transport pallet vehicles. This truck could then drive fully automatically to its destination, for example, to a logistics hub.

From the logistics hub, the packaged shoe could be flown directly to my front door with a parcel drone due to its low weight.

---

[5] Henning, Klaus: Kybernetische Verfahren der Ingenieurwissenschaften (Cybernetic Procedures in Engineering Sciences). Mainz, Aachen 1986.

The customer finally receives both the virtual and the real shoe. They can then decide whether the virtual shoe remains coupled with the supplier or whether they want to couple it with their other shoes, with the shoes of other people, or with an orthopedic doctor who will analyze the collected data.

This scenario has already been implemented in many of its elements. There is still a lack of consistency throughout the entire value chain. But that's just a matter of a few years.

> **In the future, many products will come to me by themselves.**

Services are therefore becoming increasingly complex and diverse to the benefit of the customer. Many of these service models would be inconceivable without digital networks and integrated AI processes. Many things are generated fully automatically. This includes trend analyses. It will be easy to see the best time to visit a gym, supermarket, or restaurant. With one click you can see the visitor statistics depending on the time.

The entire communication between customer and supplier is increasingly established via the internet. Customer and supplier no longer need to meet physically. This makes it even important that a personal conversation is possible if the customer wishes to do so. In the case of complaints and delivery defects as well as maintenance, direct communication from person to person must be enabled efficiently.

In summary:

- Data and information about everything are available everywhere and at any time.
- In order to obtain data and information, a physical presence at a certain place at a certain time is less and less necessary.
- The value chains of product creation, logistics, and services will be permeated with intelligent digital shadows in all elements, at all locations, in the overall processes and in every small detail.
- In addition to fully automated communication chains, it is a competitive advantage if parallel communication channels for direct human-to-human communication exist as a hybrid intelligence.

12

The Ethical and Legal Implications

Contents
The Machine Becomes a Legal Entity 100
What Ethical Standards Do We Want? 102

Digital shadows will permeate to a considerable extent of all value-added processes of products and services, both large and small. In 15–30 years, these AI systems will be widespread and highly mature, and will characterize wide areas of industry, public and private life. This will fundamentally change the way we think and act. Competence development, educational pathways, legal systems, and public order systems must be rethought. Moral and ethical issues need to be addressed.

Artificial intelligence will take over many work processes. This also applies to public administrations, law firms, courts, and notaries.

The creation of new documents from existing ones or the retrieval of judgments in connection with a legal problem is no problem for a suitably trained AI system.

> **The AI machine will be able to prepare the legal contract document to a wide variety of cases.**

However, the AI system also bears its own responsibility for what it does. The many application examples of AI systems—autonomous cars, autonomous forklifts, and care robots—raise many new questions for insurances, laws, and ethics.

It is no problem for a "dementia robot" to be a meaningful digital shadow in the spaces of its physical partner. As a robot it will accompany the patient like a familiar pet and will not become annoyed by having to repeat the same thing 50 times. Yes, it will also adapt to the patient's "world view" and modify its behavior to the degree of dementia, train the patient with questions, but above all ensure that the dwindling memory capacity is replaced by the machine's own "brain" where necessary. The demented person thereby experiences a kind of expansion of their own brain and can compensate for deficits.

> **The dementia robot can have a patience that the humans are incapable of.**

My late mother-in-law would have used it with enthusiasm. She lived alone in her house and to fight the progression of dementia she had gotten used to writing everything on sticky notes and hanging them on the walls. The whole house was full of notes like "Lock the door tonight." "I want to call my sister." "Is the stove off?" "Did I lock the front door?" It would be easy for an automated caretaker to manage all these ideas in place of sticky notes. Instead of writing them down, the caretaker robot could simply listen to what the patient was saying—reminding her of certain tasks (like asking if she locked the door) when appropriate. If necessary, the robot could simply lock the door itself. For tasks beyond the physical capabilities of the robot, a neighbor could be called to help.

However, this raises questions about the "personality" of AI systems and the responsibility for misconduct and accidents caused by them. Who takes over liability in the event of a fault?

The Machine Becomes a Legal Entity

In an accident with a fully automatic vehicle of a taxi company, a cyclist was run over.[1] To be on the safe side, the vehicle also had a driver who was supposed to monitor the journey. The cyclist died as a result of the accident. In the court case, the driver was acquitted because the vehicle's cameras could prove that the driver had no chance to see the cyclist.

---

[1] Tempe police chief says early probe shows no fault by Uber. https://www.sfchronicle.com/business/article/Exclusive-Tempe-police-chief-says-early-probe-12765481.php, accessed in April 2020.

And the vehicle? Was it a design fault? Perhaps the camera that discovered the cyclist was not part of the vehicle's AI system at all, but a "digital fool" who only takes pictures? In this case it would be a question of producer liability. The general question is: When is an automatic vehicle liable for its misconduct?

However, if the camera systems are part of the vehicle's integral AI system, which is responsible for all driving situations and driving states, then the vehicle must adhere. The AI system needs its own insurance.[2] The developer has no chance to predict all situations in which an intelligent AI machine must act and determine the expected action. It makes sense for an intelligent AI machine to think for itself and act on its own responsibility. But it must also become its own legal entity. It is therefore necessary to redesign legal systems both nationally and internationally.

> Autonomous cars, forklifts, and healthcare robots raise new questions for insurance companies, laws, and ethics. Intelligent machines with their own consciousness must become their own legal entities accountable to corresponding laws.

This also applies to the much-discussed accident dilemma. The imaginary accident situation looks like this: A fully AI-controlled autonomous vehicle finds itself in a critical situation in which it can only decide to hit an 80-year-old person or a six-year-old child. What should it do?

The crucial thing about this situation is that a human driver is in shock in such situations of fractions of a second and in most cases cannot be held responsible for having run over the child in order to save the 80-year-old person.

In the case of the AI machine, however, the ethical (and legal) question of the decision arises because the AI system will not be in shock. And if it is capable of reacting fest enough, then it is also responsible for its decision. But what ethics and what legal standard do we give the AI machine in its "driving school"?

> There are many ethical and legal questions relating to actions over which the humans cannot decide because they are incapable of it.

---

[2] Wildemann, Horst: The Limits of Artificial Intelligence. https://www.welt.de/wirtschaft/bilanz/article188571271/Maschinelles-Lernen-Die-Grenzen-kuenstlicher-Intelligenz.html=- proudly presents, accessed in April 2020.

The fact that AI systems allow us to observe actions that we as humans are not able to do has even more far-reaching consequences.

Similar to the problem of non-reproducible weld seams (see Chap. 3), we can use AI algorithms to predict impending crimes with a very high probability in order to carry out preventive burglary diagnostics. However, this raises the question of whether the *intent* to commit crime is a punishable offense.

This question has, of course, always played a role in legal assessments. But the scale and accuracy of such crime prediction reach a completely different dimension through AI systems. Therefore, the question of preventative police action must be clarified from a legal point of view.

> AI agents can also be used for preventive police operations. Is mere *intent* a punishable offense?

There will be many cases in which it makes sense to use a digital "crime shadow," i.e. an AI crime agent, instead of the real police officer or detective. But this too will have to become a separate legal entity. Here it will come to a hybrid intelligence of human and AI criminal agents.

These examples make it clear that in the long run there is no way around the fact that AI algorithms need ethical standards. But these must also be agreed internationally.

What Ethical Standards Do We Want?

The system of order introduced in China of awarding or subtracting points, respectively, for either acceptable or unacceptable public behavior, will probably not become the European or American way. But we will certainly have to fundamentally rethink, for example, the way we monitor traffic.

This topic has been a national topic in Germany since autumn 2018. The cornerstone paper of the Federal Government on Artificial Intelligence says, among other things:[3,4]

---

[3] https://ec.europa.eu/knowledge4policy/publication/germany-artificial-intelligence-strategy_en, accessed in April 2020.
[4] https://www.bmwi.de/Redaktion/EN/Pressemitteilungen/2018/20181116-federal-government-adopts-artificial-intelligence-strategy.html, accessed in April 2020.

- It is important to promote the responsible and public interest use of artificial intelligence in cooperation with science, industry, the state and civil society.
- We need a Europe and American response to data-based business models. We need to find new ways to create data-based value that reflects our economic, moral, and social structure.
- The use of artificial intelligence for human beings must be at the center of attention—at the personal, individual, and social level.
- It must be examined whether ethical and legal limits to the use of artificial intelligence fit and whether the regulatory framework for a high degree of legal certainty needs to be further developed.

Similarly, the American National Artificial Intelligence Research and Development Strategy Plan states, among others (strategies 2–4):[5]

- The aim is to develop the best methods for collaboration between humans and AI systems.
- It is important to understand and address the ethical, legal, and societal implications.
- Safety and security of AI systems must be ensured by improving explainability, transparency, and trust.

In a report of February 2017, the European Parliament calls for the long-term "creation of a special legal status for the most autonomous robots as 'electric persons'" in order to differentiate them from humans.[6]

The discussion is in full swing. And that's a good thing. New legal systems and new social order systems do not emerge overnight. Like the discussion about genetic engineering, such process takes time. We still have that time although not more than 15–30 years.

Under the influence of AI systems with their own consciousness, the political regulatory systems must be redesigned.

---

[5] The National Artificial Intelligence Research and Development Strategic Plan: 2019 update. https://www.whitehouse.gov/wp-content/uploads/2019/06/National-AI-Research-and-Development-Strategic-Plan-2019-Update-June-2019.pdf, accessed in April 2020.
[6] Committee on Legal Affairs of the European Parliament, 2015/2103(INL) (Jan. 27th, 2017): Recommendation to the Civil Law Rules on Robotics, Liability, AC to AF. https://www.politico.eu/article/europe-divided-over-robot-ai-artificial-intelligence-personhood/, accessed in April 2020.

It is only the beginning of the Inverse Gutenberg Revolution. During the first Gutenberg revolution, some 600 years ago, it took almost 200 years for a relative stability to reappear in the European system of order after the "revolution of reason." This time, the effects are global; and the timeline is likely to be much faster. Back then, even the 30 years from the invention of book printing to general reading ability was still fast enough to seem a serious "disruptive shock" to the status quo.

We will have to develop a new culture of borders. My generation was busy overcoming borders and abolishing borders. Even the borders of the Iron Curtain have fallen, and the Berlin Wall is gone.

However, we notice that the approach of boundlessness is reaching its limits. For example, the high global transparency of data has improved communication between people to such an extent that smuggling gangs can optimize refugee flows using the internet. However, the best escape routes, border crossings, and weather data can also be easily determined. Without such networking through smartphones and social media, the mass flight to Europe in the years 2014–2016 would not have been possible. The same applies to refugee flows from South to North America.

The value of a border for the adequate protection of people and national or regional identity must be rediscovered. In most areas we do not know yet what new meaningful boundaries look like in the age of digital transformation with artificial intelligence.

However, if this fails, the number of physical boundaries for the protection or defense of people worldwide will grow again—and that is certainly not the appropriate answer to the worldwide transparency of data and processes.

The discussion about the ethical standards for artificial intelligence is in full swing. Terms like robocalypse make the rounds. There is talk of the curse and blessing of artificial intelligence. While some plan[7] the robot factories in space, others warn against a totalitarian surveillance state[8] and a deadly arms race with autonomous weapons. But there are also worldwide initiatives that work to ensure that artificial intelligence serves the well-being of people (Tegmark, 2017).

Many questions remain unanswered. What does the legal entity of an artificial intelligence system look like? How can I communicate standards and values if the influencing factors for such systems are not known? In dealing with artificial intelligence, we still lack the concepts to regulate it.

---

[7] Space Tango start-up is planning a robot factory in space (2020). https://spacetango.com/, accessed in April 2020.
[8] https://en.wikipedia.org/wiki/Max_Tegmark, accessed in April 2020.

But the current rating system for likes and clicks, in which like-minded people can win worldwide majorities online in a matter of minutes, is certainly not a forward-looking concept. The science journalist Ranga Yogeshwar[9] argues that the resulting "filter bubbles" are poison for a democratic society.

Perhaps the first thing we should remember is the writer and biochemist Isaac Asimov, who demanded "basic rules" for the use of robots back in 1942 (Asimov, 2008):

1. A robot shall not (knowingly) injure any human being or (knowingly) allow any human being to be harmed by inaction.
2. A robot must obey the commands given to it by a human - unless such a command would collide with rule one.
3. A robot must protect its existence if this protection does not collide with rule one or two.

> **We should do better than our ancestors in the Gutenberg era. We should design the process of change triggered by artificial intelligence proactively and keeping with our social values. Before others do so irresponsibly.**

---

[9] https://en.wikipedia.org/wiki/Ranga_Yogeshwar, accessed in April 2020.

13

Guidelines for the Necessary Redesign of Our Regulatory Systems in Industry and Society

Contents
The Power of Shaping the Future.. 108
The Value of Trust.. 112
The Value of Agility... 113
The Value of Mindfulness... 115
Integrated Design of Core Processes... 116

We have now seen the potential opportunities and risks that might arise from the worldwide use of AI systems with their own consciousness. Most of these developments are unstoppable, at least as much so as the upheavals in people's lives caused by the advent of mass printing 600 years ago.

> It's about regaining the courage to embrace the power of creating and designing.

We have seen that this is a unique opportunity to redesign all aspects of our private, professional, and social lives from the bottom up. But it can also fail if we take too long, and the global dynamics are determined by others. Societies or certain groups of people could approach the question of the use of artificial intelligence with a different view of the world and a different understanding of values. It is obvious that artificial intelligence and digital transformation represent an opportunity that needs to be handled responsibly. What are we supposed to do now?

The Power of Shaping the Future

I recall an interesting interaction I had with a friend of mine who is a 16-year-old student.[1] He was tasked with writing a paper on the neuroevolution of software programs with artificial intelligence. The paper was due in 3 weeks. He decided to build a small machine as the basis of his work. He programmed a small "AlphaGo Zero,"[2] a gaming machine solely provided with the rules of the game that must learn everything else itself. Of course, he equipped his program with a simple neural net, as you could already find in the textbooks of 30 years ago.

The foundation of his program were 14 individual virtual battle tanks which entered a competition in pairs, competing to see which of the two were better at hitting simulated targets. After the first two, the next two took their shots, then the third pair, etc., so that every tank eventually competes against each other. After the first round is complete, the next round of learning starts, we would call this the next "generation" of the learning cycle. While he was writing his paper, the small AI machine ran day and night on his laptop, because although the laptop was powerful, it turned out to be a little overwhelmed by the task.

By the time the student had finished his paper, the learning process of the AI simulation was advanced enough that it could show results: The simulation had actually learned to develop itself further, solely by trial and error, i.e. with reinforcement learning, the reform pedagogical approach.

He also learned that it is crucial to set the right rules at the beginning of the machine's learning process and how sensitive the activation functions are. These are used in the artificial "neurons" to determine which criteria should be used to align the game with its objectives. For this he applied the principle of mutation and selection in the program but had to start from scratch several times. The general conditions were such that the game could not develop, and for example, all the tanks turned in circles and wouldn't do anything else because they were slaved to the initial search function.

We discussed this and he said that the reward functions for the system would be comparable to the ten commandments of the Bible, which—as rules—should guide people in their personal development.

I was amazed. But I was even more amazed when I learned that the computer science teacher to whom he had to deliver the work was very skeptical

---

[1] Roth, Lukas: Self-learning algorithms through neuroevolution. Cologne 2019. https://henning4future.com/en/contact/, accessed in April 2020.
[2] https://de.wikipedia.org/wiki/AlphaZero, accessed in April 2020.

about the subject. I realized that my friend, the student, became the teacher of his teacher. This presupposes that the teacher has the insight and willingness to recognize his student as a teacher. In this sense, a student also has creative power in his "profession."

> **Where we work and live, we can be "powerful" and shape the future.**

It is necessary to understand this so that we do not unconsciously wield power. My creative power transferred in my profession is my contribution to the further development of the world—be it the digital transformation, or the care of a person, or the repair of shoes.

The theologian Romano Guardini says: "The power of creation is a good gift of God - through it I am a co-creator of this world" (Guardini, 1957).[3]

In this sense we could then also say that artificial intelligence is a gift of God, given that we also exercise the power to create and set the right framework conditions.

> **Artificial intelligence is a gift of God.**

The question is whether this still makes sense given the complexity and dynamism of the world (Henning, 1993). World development is characterized by the fact that contradictions are growing (Fig. 13.1). Our ability to act has much improved—we understand, and we can do more thanks to technological advancements. At the same time, our ability to perceive is constantly increasing. We can perceive and observe more and more processes in this world.

The perceptual space, i.e. everything we *could* perceive, grows much more strongly than our *ability* to perceive. That means, the difference between "what we *should* perceive" and "what we *can* perceive" grows disproportionately. Even for the sake of pure self-protection, we are dependent on not perceiving everything that we could perceive. The filtering strategy of Homo zappiens is especially helpful for this purpose.

In other words, although our ability to act is getting better and better, the amounts of things we perceive but cannot manage to act upon increase disproportionately. And the perceptual space grows disproportionately in comparison to the perceptive ability.

---

[3] https://en.wikipedia.org/wiki/Romano_Guardini, accessed in April 2020.

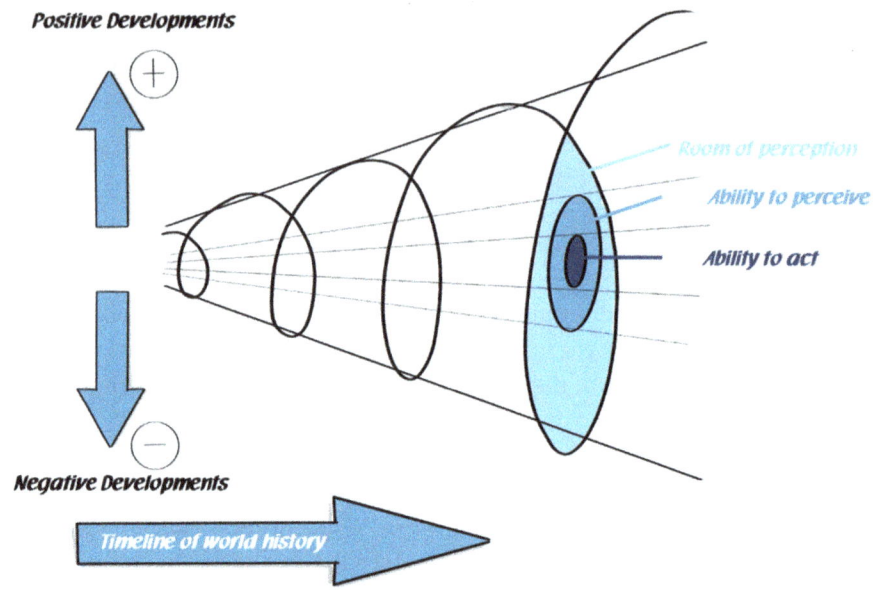

Fig. 13.1 Growing power and powerlessness

The power of shaping the world around us grows, while at the same time the powerlessness grows.

What are we to do?

In order to shape this future responsibly, agile structures and forms of work are needed in which people courageously proceed, try out new things, and launch new products, services, forms of work, and ways of life based on trusting cooperation despite all the complexity and dynamism that surrounds us.

> **Anyone who does not try to shape our future on a small or large scale, being agile, fast, flexible and full of trust, has not understood the signs of the times.**

The shaping of our future—with artificial intelligence as the gamechanger—presupposes, that we can deal with the growing dynaxity (Valtinat, Rick, & Henning, 2006).[4] A distinction is made between the dynaxity static, dynamic, turbulent, and chaotic zones. At the latest in the turbulent zone, a high level of factual as well as emotional competence is required for this.

---

[4] https://en.wikipedia.org/wiki/Dynaxity https://henning4future.com/en/dynaxity/, accessed in April 2020.

With increasing dynaxity, we humans tend to retreat into the "cozy corner" where we can all love one another and where we repress the truth. "It's not going to be that bad with artificial intelligence." "Oh, no, you're exaggerating, it's the scientists with their airy ideas." or "I won't live to see it, anyway, why should I care." Unfortunately, there are also remarks like: "I don't want and need a smartphone."

This attitude usually tips over when reality catches up and the mood changes. Now only facts, data, and the bitter reality count. If we do not succeed in designing the systems of artificial intelligence proactively in a responsible manner, the bitter reality will catch up with us faster than we would like. Then, Europe and the Americas could be overwhelmed by Chinese procedures and value systems for the use of artificial intelligence, for example.

The happy medium lies in the balance between truth and love, honest and emotional, full of trust as well as with a lot of control, capable of conflict and reconciling. The increasing complexity and dynamics cannot be countered by simplification, repression, nationalism, or populism.

> **The values of trust, agility, and mindfulness are key to shaping the future of digital transformation.**

It is important to find and rediscover a balance between love and truth in all areas of life. The necessary must be done. At the same time, the people concerned should be taken along in a mindful and loving manner.

It is the task of managers and responsible persons in all areas of our society, either parents in their educational role, or teachers, group leaders, social workers, doctors, managers, politicians, etc., to "live" such a path.

Trust, agility, and mindfulness are the three values that seem necessary to successfully shape the transformation process in such a way that they are all lived and implemented simultaneously.

These three values are based on decades of experience in dealing with turbulence in organizations.[5] There are certainly numerous other important values, but the exchange of experiences with those affected has shown me again and again that these three values are decisive if transformation processes of a turbulent nature are to take place successfully and with as few undesirable side effects as possible.

Let's take a closer look at these three values:

---

[5] Over many years, we have searched for success factors in change projects that we have accompanied. Backgrounds to it. In: Henning, Klaus: The art of the small solution (Henning, 2014).

The Value of Trust

At first glance, a culture of trust is nothing new. At second glance, however, that means:

Trust in hybrid intelligence, not only between people, but also between organizations, and trust in what happens between machines. This tension also describes the trust between the three legal categories—people, organizations, machines. They must be reassigned not only ethically, but also legally. Binding rules for dealing with each other must be agreed upon.

Between people and organizations, we are familiar with that. People are also legal entities. Organizations also exist as legal entities—as a corporation, a limited liability company, or an association. Machines with their own consciousness have yet to become legal entities. The new legal category of AI systems with own consciousness will appear. Just as the legal coexistence of people and organizations sometimes works well and other times not so well, the same will be the case in the triangular relationship between man, organization, and machine.

It is not only about the legal aspect, but about all aspects of the varied coexistence between people, organizations, and machines with their own consciousness, because from a systemic point of view they are all "living beings."

In particular, the machines with own consciousness must be educated. Similar to a dog that must go to a training before it can become an avalanche search dog, for example. Likewise, the fully autonomous car, the robot on four wheels, must pass through a driving school where it learns in the company of other fully autonomous cars before it gets released onto the road or onto the public. You don't leave a little kid out on the street alone either. But at an agreed time of competence development and learning processes, the machine will grow up with its own consciousness and can and must become responsible for itself. Such a legal entity may need to be capitalized for liability purposes or appropriate insurance schemes may be developed.

If such a machine is a legal entity, it will also be subject to criminal law. It cannot be ruled out that among the many autonomous cars a few will develop into rogue drivers through experience and exchange of experience. If they are intensively oriented towards rogue human drivers and are in an international network of AI rogues, this can happen quickly.

In such cases there will certainly have to be criminal charges against autonomous AI machines, just as I can sue an automobile company for violating the emission values. And a trial against an AI car will then also result in the possible suspension of their license or imposition of the "death penalty," i.e.

withdrawing the vehicle from circulation. The moral problem of the death penalty will not arise with artificial intelligence machines. Unless the life or work of many people is threatened by switching off an intelligent machine. Just as the closure of a company ordered by a court can deprive many people of their livelihood.

Building this new culture of trust between people, organizations, and machines is quite a big task.

> We need a culture of trust—vertical and horizontal—between people, machines, and organizations.

Added to this are the dimensions of vertical and horizontal trust. Many social systems in the private and professional sectors are characterized by strict hierarchies and departments. You don't trust your neighbor. You don't trust the colleague in the neighboring office. You don't trust the customer. You don't trust the supplier.

The same thing happens between the hierarchies. You don't trust the bosses. You don't trust the staff. You don't trust the kids. You don't trust the parents.

The digital transformation with artificial intelligence, however, will link everything with everything. Things become transparent across all hierarchies along the entire value chain around the world.

In this context, it becomes clear what is meant by the statement: A culture of trust— vertically and horizontally—between people, machines, and organizations is necessary.

The Value of Agility

It is similar with the second value, agility (Weiler, Savelsberg, & Dorndorf, 2018). What does agility mean today?

The term has acquired a new quality through experience in the design of very complex software systems. In 2002, as part of a consulting project, we extended it to a design principle for change processes in systems with high turbulence[6]—according to the Agile Manifesto for Software Development:[7]

---

[6] https://henning4future.com/en/agile-manifesto/, accessed in April 2020.
[7] According to the Agile Manifesto of Software Development—https://henning4future.com/en/agile-manifesto/, accessed in April 2020.

- We focus on people and their interactions—with a focus on ongoing processes.
- Individuals and interactions are more important to us than processes and tools.
- Running processes are more important to us than extensive documentation.
- Cooperation with the customer is more important to us than contract negotiations.
- It is more important for us to be able to react to changes than to pursue a plan.

Therefore, although the second things have their value, we assign a higher value to the first mentioned things.

> **It needs agility as a design principle for all areas of life and work in all organizations of public and private life.**

This approach fits well into Stafford Beer's[8] understanding of organization (Borowski, 2011). He has derived an approach from the structures of the human brain with the "Viable System Model." It explains how organizational principles of the brain can be transferred into the organizational cooperation of humans. Stafford Beer distinguishes three levels: Values, principles, and "norms."

At the value level, these are the values of trust, agility, and mindfulness. The value agility is well described by the agile manifesto as described above. At the level of norms, however, something very important is at stake: the principle of small steps with small solutions (Henning, 2014).

As an organizational principle, the next step takes place based on a common view of the future by agreeing on binding deadlines and resources. However, it does not specify how far I will go. The only rule that applies is: "Run as fast and well as you (with your team) can - within the limits of resources in terms of people, time, budget, technology and within the agreed deadline."

This operational level is very close to the organizational principle of learning and experience, allowing trial and error. It resembles the operation of machines with their own consciousness. But it is also derived from the basic

---

[8] https://en.wikipedia.org/wiki/Viable_system_model, accessed in April 2020.

knowledge about the functioning of the brain. Recursive cycles of knowledge lead quickly, but in small steps, to success.

The brain always works in a team. Each layer of neural networks consists of a myriad of parallel nodes that exchange data and even attenuate each other (lateral inhibition). However, it is precisely through their work "in a team" that they quickly gain sharpness in the reproduction of the observation made.

If we apply this approach to the interaction between people, organizations, and intelligent machines, it becomes clear that it is a question of working together at eye level. For example, the administrations are concerned with interdepartmental action. In industry, it is all about the cooperation of marketing, sales, maintenance, development, and testing. What a long way still lies ahead of us to an agile cooperation…

But there are encouraging examples of how companies and states are becoming agile. The company Spotify, which has a completely agile structure, should once again be mentioned as an example. It consists of extremely autonomous agile teams grouped into "tribes" that have enabled rapid and enormous growth.[9]

And there is the example of Estonia with its 1.3 million inhabitants. It has become a model for the digitalization of public administration worldwide.[10] A great deal can be handled on the internet. This allows Estonian citizens to sign digitally without any problems; they can manage their medical records themselves. An entrepreneur can also handle the complete establishment of a company with all legal transactions on the internet.

This concept of "e-residency" could be imported into other countries and to mega-cities. We don't have to reinvent everything ourselves.

The Value of Mindfulness

Thirdly, it requires a high emotional ability of awareness. Of the three values for digital transformation, the value of mindfulness is certainly the most difficult to implement.

This value requires the ability to relax and learn to endure the complexity and dynamics of my environment without repressing it, without becoming

---

[9] Spotify Engineering Culture, 27.02.2017. https://www.youtube.com/watch?v=4GK1NDTWbkY, accessed in April 2020.
[10] Adi Gaskell (Jun 23, 2017): How Estonia Became The Digital Leader Of Europe https://www.forbes.com/sites/adigaskell/2017/06/23/how-estonia-became-the-digital-leaders-of-europe/#c0f3b12256da/, accessed in April 2020.

angry or depressed. But above all it's about not falling into a panic of action, according to the motto "Nobody but me can do it" (Henning, 2015).[11]

Objective wide and emotional perception with a big heart requires relaxed perception. It takes more than that:

Managing systems with high dynaxity requires intelligence, sensitivity, and "laziness"; the latter in the sense of relaxed perception, the ability to reflect and anchor oneself in one's own sense of life—relaxed, resting in oneself, finding oneself, meditating, praying.

> **What is needed is the art of mindfulness and awareness, which does not suppress growing dynaxity. It rather perceives it in a relaxed way, endures it, and renounces the reflex of action: "Now something has to be done very quickly!"**

If the three values of trust, agility, and mindfulness are brought together and if we begin to implement them together in a goal-oriented manner, there is a good chance that the digital transformation with AI systems will succeed.

Integrated Design of Core Processes

It is important to bring three things in the direction of the desired goals and to design the framework conditions between people, organizations, and machines, i.e. hybrid intelligence (Fig. 13.2) (Hanna, 2013; Henning & Meinecke, 2017):

- The task-core processes of people, organizations, and AI systems must be oriented towards a jointly agreed and lived goal, namely the purpose of an organization.
- The attitudes, ethical values, and behavioral culture for all three "systems"— i.e. the individual core process—must also be oriented towards the common goal. This concerns not only the human being, but also the "individual" process of an organization and the individual process of an AI machine with its own consciousness.
- The Social Core Process, which constitutes the coexistence in hybrid intelligence, i.e. the coexistence of people among each other, machines among each other, the coexistence between people and machines, is—after all— the decisive factor for a successful transformation. But this also includes the

---

[11] https://henning4future.com/en/the-ego-trap-7-ways-to-ruin-your-business/, accessed in April 2020.

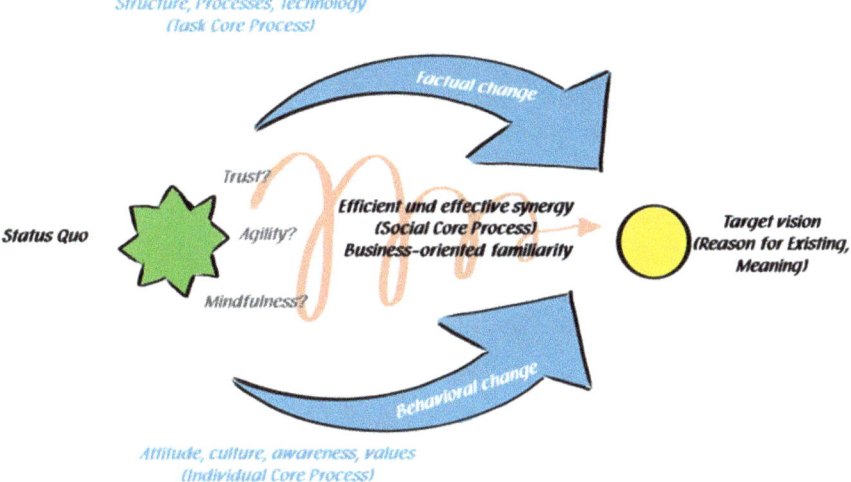

Fig. 13.2 The core processes of the OSTO system model (https://en.wikipedia.org/wiki/OSTO_System_Model; https://henning4future.com/en/osto-system-model/, accessed in April 2020.)

coexistence of the digital shadows, of all participants and their communication with each other and their respective real partners.

We should design this hybrid intelligence in a positive way:

- The system boundaries of states, public order systems, companies, and institutions of all kinds (public offices, churches, associations) will change dramatically under the influence of artificial intelligence.
- Democratic control structures—perhaps modelled after the social market economy and the German model of democracy—must find their way into the systems of artificial intelligence.

It affects all levels—the individual person, the family, the municipal structures, the companies, the authorities, the large corporations, our legal systems, and the state. That is the task of the Inverse Gutenberg Revolution. It is a century task that will hopefully not take 100 years so that we have less turbulence than our predecessors during the Gutenberg Revolution.

14

Epilogue: Does Artificial Intelligence Make God Redundant?

Yuval Harari (Harari, 2018), one of the world's most famous historians, writes in his book "21 Lessons for the twenty-first Century": "Morality does not mean 'adhering to divine commandments'. It means to reduce suffering."

But do we still need the question of God at all if we expand our consciousness through the amount of intelligent AI machines? Yuval Harari sees us on a path where humans become gods. The "Homus Deus" (Harari, 2015), the all-knowing and understanding human being, who will ultimately be able to solve all basic human questions, will emerge.

In the history of mankind, people have repeatedly tried to answer the question about God. However, this is not so easy.

The French philosopher and theologian Pierre Teilhard de Chardin shaped the term noosphere in his book "The Origin of Life" in 1950. By this he means a spiritual sphere that shapes the development of culture and individuality.

Looking towards the future, he wrote at a time when nobody thought about digital transformation by machines with artificial intelligence (Teilhard de Chardin, 2002, 2005):

> From man (from the last and highest point of this evolution of the *first degree*) ... it is about structures that are calculated, complement each other and connect with each other. Are we not dealing with an evolution that combines its forces into an entirely new kind of advance, which only became possible because this <first> evolution became aware of itself? With a *second-degree* evolution, a conscious evolution?

If one follows this thought, then one could regard the current development towards hybrid intelligence as an extension of the human brain. This is very close to the approach of Yuval Harari and the discussions about cyborgs (cybernetic organism).[1] This refers to mixed beings in which robots and AI systems become part of the same organism.

Is the human being at the end of the development of a single hybrid intelligence, regardless of whether the cooperating AI systems are in the human's environment or are even implanted as technical systems in the human being? The theologian Volker Jung writes (Jung, 2018): "Digitalization deifies humans and robs him of himself at the same time. What remains unclear in Harari's construction is what room for scope of action really exists."

If one looks at the world from the modules of matter, energy, and information, then the decisive room for scope of action is handling of information as the decisive shaping force. This approach is described in the Bible at the beginning of the Gospel of John where the emergence of the world begins with the sentence: "In the beginning was the word." For the word "logos" used in Greek, however, the word "information" fits much better. It is helpful to regard the word "information" in the scientific sense as a measure of chaos and order (Henning, 1980).

From this point of view, God dwells in the diversity of information within the human themselves. "And the word became flesh and dwelt among us." The thoughts of God, also known as the "Holy Spirit," then dwell in the human brain. But since we now know that our consciousness—at least as far as P-consciousness is concerned— is also distributed in the body, God dwells in everything that can develop its own consciousness. In this sense, there is a hybrid intelligence between God and the human.

If divine presence "dwells" in the consciousness of the human, then it could also dwell in the consciousness of animals and in the consciousness of machines.

If we follow the idea of the second-degree evolution by Teilhard de Chardin, then God would also dwell in the objects with their own consciousness. Then, it would also be possible for the "Holy Spirit" to spread in the networks of objects with their own consciousness. Hybrid intelligence between humans and AI systems would then also be an interesting partner at the level of faith.

When it comes to "reducing suffering" through globally agreed ethical values, i.e. a new morality, then perhaps it would not be entirely wrong to think about a hybrid intelligence of humans and systems of artificial intelligence in

---

[1] https://en.wikipedia.org/wiki/Cyborg, accessed in April 2020.

communion with God. After all, the Gospel of John says that the Son of God, Jesus Christ, had placed himself in the midst of the suffering of the world.

We would then have found an approach to reducing the suffering of this world in the hybrid intelligence of the human, machine, and God in a kind of shared enlarged consciousness.

About the Author

Prof. Dr.-Ing. Klaus Henning studied electrical engineering and political science, received his doctorate on human-machine systems, and his professorial certificate in Entropy in Systems Theory.

For 25 years, he headed the cybernetics lab of the RWTH Aachen University. For years, Prof. Henning was a member of the Presidium of the VDI, Vice Rector for Finance at RWTH Aachen University, and Dean of the Faculty of Mechanical Engineering at RWTH Aachen University. For over 10 years, he was a member of the University Council of Saarland University and Chairman of the supervisory board of Xenium A.G., Munich, member of the scientific advisory board of the federal CDU Economic Council and numerous other advisory boards in the academic and industrial environment.

Today, he works as independent consultant and as senior partner of umlaut transformation GmbH, a company of umlaut S.E. He is member of the board of the Institute for Corporate Cybernetics (IfU e.V.) at RWTH Aachen University.

Currently, most of the customers he serves—mostly at board and department head level—come from the IT industry, university hospitals and the supplier industry of mechanical and plant engineering, the automotive industry, the aerospace industry and logistics.

He was scientific coordinator of the Future Dialogue of the German Chancellor Angela Merkel (2011–2012).

He summarized many of his experiences in a book "Die Kunst der kleinen Lösung" (The art of the small solution)—how people and companies master the complexity (Henning 2014).

Bibliography

Asimov, I. (2008). *Foundation*. New York: Batam Books.
Bianchini, C., Osthoff, C., Souza, P., Ferreira, R. (2020). High performance computing systems. In *19th symposium, WSCAD 2018*, São Paulo, Brazil, October 1–3, 2018, Revised Selected Papers. Zuerich: Springer.
Borowski, E. (2011). *Agiles Vorgehensmodell zum Management komplexer Produktionsanläufe mechatronischer Produkte in Unternehmen mit mittelständischen Strukturen (Agile process model for managing complex production start-ups of mechatronic products)*. Düsseldorf: VDI. Reihe 16, Technik und Wirtschaft.
Brandt, D. (2003). *Human-centered system design*. 20 Case Reports. Aachener Reihe Mensch und Technik, Bd. 42.: Verlag Mainz.
Campe, R. (2017). *What's App, Mama?: Warum wir Teenies den ganzen Tag online sind - und warum das okay ist!* Hamburg: Eden Books.
Dawkins, R. (1989). *The selfish gene* (2nd ed.). Oxford: Oxford University Press.
Eisenstein, E. (2009). *The Printing Press as an Agent of Change - Communications and cultural transformations in early-modern Europe*. New York: Cambridge University Press.
Gabriel, M. (2018). *Der Sinn des Denkens (The meaning of thinking)*. Berlin: Ullstein.
Guardini, R. (1957). *Die Macht - Versuch einer Wegweisung (The power - Attempt to show the way)*. Würzburg: Werkbund.
Haberstroh, M. (2014). *Prospektive Analyse sozio-technischer Innovationen – Die elektronische Kopplung von Lkw auf Bundesautobahnen (Prospective analysis of socio-technical innovations - The electronic coupling of trucks on federal motorways)*. Marburg: Tectum.
Hanna, D. P. (1988). *Designing organizations for high performance*. Reading, MA: Addison-Wesley.

Hanna, D. P. (2013). *The organizational survival code: Seven capabilities to get the results you want*. Mapleton, UT: Hanaoka Publishing.

Harari, Y. N. (2015). *Homo Deus - A brief history of tomorrow*. London: Penguin Random House.

Harari, Y. N. (2018). *21 lessons for the 21st century*. New York: Penguin Random House.

Heide, A. (2004). *Ursachenanalyse und Bewertung der Verantwortung bei Funktionsstörungen von softwaregesteuerten Komponenten im Maschinenbau* (*Cause analysis and assessment of responsibility in case of malfunctions of software-controlled mechanical components*). Düsseldorf: VDI.

Heide, A., & Henning, K. (2006). The "cognitive car": A roadmap for research issues in the automotive sector. *Annual Reviews in Control, 30*(1), 197–203.

Henning, K. (1980). *Entropie in der Systemtheorie* (*Entropy in systems theory*). Habilitation thesis, RWTH Aachen, Aachen. https://henning4future.com

Henning, K. (1993). *Spuren im Chaos. Christliche Orientierungspunkte* (*Tracks in the chaos. Christian landmarks*). München: Olzog. https://henning4future.com/spuren-im-chaos-christliche-orientierungspunkte-in-einer-komplexen-welt/

Henning, K. (2014). *Die Kunst der kleinen Lösung. Wie Menschen und Unternehmen die Komplexität meistern* (*The art of the small solution. How people and companies master complexity*). Hamburg: Murmann.

Henning, K. (2018). How artificial intelligence changes the world. In A. Karafillidis & R. Weidner (Eds.), *Developing support technologies - integrating multiple perspectives to create assistance that people really want* (pp. 277–284). Berlin: Springer.

Henning, K., & Kutscha, S. (1994). *Informatik im Maschinenbau*. Berlin: Springer.

Henning, K., & Meinecke, M. (2017). Das OSTO-Modell für Organisationsentwicklung und die Kunst der kleinen Lösung. In A. Deister (Ed.), *Krankenhausmanagement in Psychiatrie und Psychotherapie*. Berlin: MVV.

Henning, K., Oertel, R., & Isenhardt, I. (2003). *Wissen – Innovation – Netzwerke. Wege zur Zukunftsfähigkeit* (*Knowledge - Innovation - Networks. Ways to sustainability*). Berlin: Springer.

Henning, K., & Preuschoff, E. (2003). *Einsatzszenarien für Fahrerassistenzsysteme im Güterverkehr und deren Bewertung* (*Application scenarios for driver assistance systems in freight traffic and their evaluation*). Düsseldorf: VDI.

Henning, R. (2015). *Die-Ego Falle – 7 Möglichkeiten Ihr Geschäft zu ruinieren* (*The Ego Trap - 7 ways to ruin your business*). Hamburg: Murmann.

Jeschke, S., Richert, A., Hees, F., Jooß, C. (Hrsg.) (2015). *Exploring demographics. Transdisziplinäre Perspektiven zur Innovationsfähigkeit im demographischen Wandels* (*Transdisciplinary perspectives on innovative capacity in demographic change*). Berlin: Springer Spektrum.

Jung, C. G. (1995). *Gesammelte Werke* (*Collected works*). Düsseldorf: Walter Verlag.

Jung, V. (2018). *Digital Mensch bleiben* (S. 35) (*Staying digital human* (p. 35)). München: Claudius.

Kunze, R., Ramakers, R., Henning, K., & Jeschke, S. (2009). Organization and operation of electronically coupled truck platoons on german motorways. In

Intelligent Robotics and Applications, Second International Conference, ICIRA (pp. 135–146). Singapore: Springer.

Lehmann, N. (2012). *Theory of society*. Stanford: Stanford University Press.

Luhmann, N. (2012). *Theory of society*. Stanford: Stanford University Press.

Mai, K.-R. (2016). *Gutenberg: Der Mann, der die Welt veränderte* (*The man who changed the world*). Berlin: Propyläen.

Maurer, M., Gerdes, J.C., Lenz, B., Winner, H. (Hrsg.) (2015). *Autonomes Fahren. Technische, rechtliche und gesellschaftliche Aspekte* (*Autonomous driving. Technical, legal and social aspects*). Berlin: Springer.

Merkel, A. (Hrsg.). (2012). *Dialog über Deutschlands Zukunft* (*Dialogue on Germany's future*). Hamburg: Murmann.

Ramakers, R., Henning, K., Gies, S., Abel, D., Max, H. (2009). Electronically coupled truck platoons on German highways. In *Proceedings of the IEEE International Conference on Systems, Man and Cybernetics*, San Antonio, TX, USA, 11–14 October 2009, pp. 2409–2414.

Spitzer, M. (2012). *Digitale Demenz. Wie wir uns und unsere Kinder um den Verstand bringen* (*Digital dementia. How we drive ourselves and our children crazy*). München: Droemer.

Tegmark, M. (2017). *Life 3.0: Being human in the age of artificial intelligence*. New York: Alfred A. Knopf, Penguin.

Teilhard de Chardin, P. (2002). *Toward the future*. Boston: Mariner Books.

Teilhard de Chardin, P. (2005). *Die Entstehung des Menschen* (S. 117) (*The origin of man* (p. 117)). München: Beck.

Valtinat, T., Rick, U., & Henning, K. (2006). Concurrent engineering and the dynaxity approach. How to benefit from multidisciplinarity. In *ISPE International Conference on Concurrent Engineering* (pp. 488–495). Antibes.

Veen, W., & Vrakking, B. (2006). *Homo zappiens. Growing up in a digital age*. Hampshire: Ashford Colour Press.

Weidner, R., Yao, Z., Otten, B., Linnenberg, C. (2018). Support technologies for industrial production. In A. Karafillidis & R. Weidner (Eds.), *Developing support technologies - integrating multiple perspectives to create assistance that people really want* (pp. 149–156). Berlin: Springer.

Weiler, A., Savelsberg, E., & Dorndorf, U. (2018). *Agile Optimierung von Unternehmen* (*Agile optimization of companies*). Freiburg: Haufe.

Index

A
Agile methodology, 16, 32
AI agents, 102
AI political regulatory systems, 103
Air traffic control, 2
AI working conditions, 89
AlphaGo Zero machine, 23–25, 108
Amazon, 79, 81
American National Artificial Intelligence Research and Development Strategy Plan, 103
Apple, 81
Artificial consciousness, 103
Artificial intelligence (AI)
   AI systems, 99–100
   AlphaGo machine, 23–25, 108
   areas of conflict, 86
   areas of tension, 85
   augmented reality, 87
   birth, 22
   collective labor agreements, 89
   cultures, 86
   DeepMind, 24
   digital communication, 86
   digital companions, 23–24
   digital networks, 86
   digital shadows, 87
   digital transformation, 1, 85, 87
   drones, 1, 2
   environmental conditions, 29
   ethical algorithms, 71
   experiential characteristics, 33
   face recognition, 57
   as gamechanger, 2
   as gift of God, 109
   GPT2 by OpenAI, 59
   healthcare, 89
   highly qualified jobs, 88
   household, 89
   human self-determination, 90
   industrial application, 88
   invention, 22–23
   logistics, 89
   national identities, 86
   pedagogical procedures, 38–39
   platform technologies, 89
   production, 89
   simple office jobs, 88
   smartphone app, 18
   social bots, 54–55
   Super Mario (game), 28

Index

transparent society, 56–57
virtual and extended environments, 89
Augmented reality, 72–74
Automatic fine machine, vii
Automation technology, 22, 66
Autonomous cars, 10, 18, 23, 99, 101, 112
Autopilots, 3, 22
Awareness of machines, 43, 88, 115, 118

B

B2B business, 95
"Big Data", 35, 55
Bitcoins, 72–74
Blockchain, 72–74
Brockhaus (a German publisher), 15
Business model
  digital business, 80
  digital platforms, 79
  digital shadow, 79
  hotels, 80–81
  machinery and plant industry, 83–84
  outsiders penetrate traditional industries, 82–83
  taxis, 80–81
  traditional business, 81–82

C

Cables, 74–75
Collective labor agreements, 89
Communication
  "Big Data" world, 36
  digital companions, 10
  digital revolution, 3
  direct communication, 51
  images and myths, 50
  images, icons and emojis, 49–51
  picture messages, 53
  with short messages, 43
  social bot, 54–55
  WhatsApp, 50, 51
Computer-Integrated Manufacturing (CIM), 62
Computer technology, 3, 39
Consciousness of machines, 7
Copper stamping, 4
Core processes, 116–117
Corona crisis, 17, 43–44
Cryptocurrency, 73
Cultural revolution, 60

D

Dartmouth conference, 22
Data privacy, 13–14
Data protection, 12, 14, 58
Decentralized control strategy, 30
"Deep Fake App", 55
Deep learning, 37, 40
Deep neural networks, 24
Dementia robot, 100
Democratic machine control
  classic tasks, 26
  knitting machine, 26, 28
  legislatures, 27
  with network intelligence, 27
  social system, 27
  software agents, 27
  voting, 27
Digital age, 15–16, 44, 46, 76
Digital companions, 10–12, 68
Digital fools, 12, 73, 101
Digital infrastructure, 75, 76
Digital learning and working, 7, 15–17
  See also Self learning machines
Digital platforms, 77–78
Digital revolution, 3
Digital shadows, 67, 68, 72, 99
Digital system landscape
  artificial intelligence, 71

augmented reality, 72–74
bitcoins, 72–74
blockchain, 72–74
cables, 74–75
digital platforms, 77–78
emerging second internet, 76–77
globally branched network, 71
intelligent digital shadow, 72
privacy, 75–76
radio masts, 74–75
satellites, 74–75
security, 75–76
services sector, 72
smartphones and notebooks, 71
voice machines, 72–74
wires, 74–75
Digital transformation, 119
and AI, 1, 4
digital networking, 6
disruptive innovations, 2
Disruptive innovation, 2
Divine presence, 120
Driverless trucks, 63–65
Drone systems, 1
Dynaxity, 55, 110, 111, 116

E

"Eliza" (chatbot), 22
Encyclopedia, 15–16
Energy revolution, 4
Ethical algorithms, 59
Ethical boundaries, 57–60
Ethical implications
artificial intelligence, 104
basic rules, 105
democratic society, 105
disruptive shock, 104
electric persons, 103
internet, 104
legal systems, 103
physical boundaries, 104
"revolution of reason", 104
social order systems, 103
Ethical standards, *see* Ethical implications

F

Facial recognition, 57–59, 97
Federal Government on Artificial Intelligence, 102–103
5G technology, 3, 40
"Floating" letter, 5
"Frontal teaching", 24, 38

G

Gamechanger, *see* Artificial intelligence (AI)
Generation Y, 47, 48
Generation Z, 48
Google Maps, 13, 18, 56
Gutenberg Revolution, 4, 117

H

"Health 4.0", 18
High performance computing systems, 40
"Holy Spirit", 120
Homo zappiens
"blurred looking", 56
global-regional, 44
and Homo sapiens, 44, 45, 52
linear television, 49
networking, 45
parallel and non-linear, 45
permanently multitasking, 45
at school, 46–47
skills, 44
"Homus Deus", 119
HOT approach, 61
Hotspots in service industry, *see* Smartphones

Human-machine interaction, 25, 26, 32, 33, 61, 65–68
Humans and robots, 32, 93
Hybrid intelligence, 120, 121
  automatic refueling, 64
  automatic truck convoys, 62–63
  case law, 63
  cross-border transport, 63
  economic issues, 65
  human factor, 64
  hypothesis, 61
  knowledge management, 62
  legal and transport policy, 63
  legal regulations, 64
  man–machine interaction, 65–67
  modes of transportation, 64
  organization, 61
  protection of trucks, 65
  technical innovation, 62
  technologies, 62

I

Icon communication, 45, 49–51
Images and myths, 50, 53, 56
Industrial revolution, 3, 4
"Industry 4.0", 18
Inegrated design, 116–117
Intelligent animals, 34–35
Intelligent development chain, 97
Intelligent exchange, 35
Intelligent machines
  agile methodology, 32
  "centauric intelligence", 70
  death penalty, 55
  and humans, 61
  hybrid intelligence, 92
  interaction, 115
  and networks, 7
Intelligent maintenance chain, 92
Intelligent objects, 11, 66, 70, 71, 77, 83, 88
Intelligent products, 94

Intelligent systems, 37, 93
Internet access, 21
Invention of letterpress, 4, 5
Inverse Gutenberg Revolution, 51, 60, 61, 104, 117

J

Job gamechanger, *see* Artificial intelligence (AI)

L

Legal implications
  camera systems, 101
  crime shadow, 102
  digital fool, 101
  legal assessments, 102
  police action, 102
  taxi company, 100
"Logic of failure", 51
"Logistics as a Service" (LaaS), 95–96
Low Earth Orbit (LEO), 76

M

Machines with ego-consciousness, 37
Man–machine interaction, 65–67
Manufacturer's onboard navigation system, 13
Mass data, 14, 60
Maturity of AI systems, 88
Medical monitoring, 16
"Meme", 53, 54
Microsatellites, 76
"Mobility 4.0", 18

N

"Near Orbit Satellites", 76
Nerve cell, vi, 40
Neural network, v, 11, 22, 24, 28, 40, 60, 115

Neuroevolution, 108
New insurance systems, 96
Non-linear learning, 45

O

Omnipresent, 11, 67, 69, 70, 87
Operative consciousness, 34
"The Origin of Life", 119
OSTO System model, 117

P

Paperback encyclopedia, 14–15
Penetration of AI systems, 88
Phenomenological consciousness (P-consciousness), 37, 120
"Picture messages", 53
Power of creation
   AI systems, 107
   digital transformation, 107
   integrated design, 116–117
   shaping
      artificial intelligence, 109, 110
      artificial "neurons", 108
      attitude, 111
      complexity and dynamics, 111
      digital transformation, 109
      filtering strategy, 109
      growing power and powerlessness, 109, 110
      learning cycle, 108
      neuroevolution, 108
      simulation, 108
      values, 111
   societies/certain groups, 107
   value of agility, 113–115
   value of mindfulness, 115–116
   value of trust, 112–113
Product development process
   central AI system, 94
   democratized control, 93
   digital shadow, 92, 94
   exoskeletons, 93
   humans and intelligent machines, 92
   integrated planning, 92
   LaaS, 95–96
   near orbit satellite technology, 94
   ramp-up processes, 93
   residential building, 92
   software-controlled component, 94
   technology, 92
   transparent, 94
   urban development projects, 92
   vehicle control systems, 94
Programming language, 21, 22
Protection of the Constitution of Bavaria/Germany, 61

R

Ramp-up processes, 93
Reducing suffering, 120
Reinforcement learning, 24, 25, 38, 39, 88, 108
RoboCup Logistics League, 29
Robotic teamwork, 29–32
Robots
   autonomous pallet control, 31
   "basic rules" for use, 105
   for dementia patients, 17
   "dementia robot", 100
   "Eliza", 22
   factories, 104
   and humans, 32
   in RoboCup Logistics League, 29, 30
   self-learning, 32
   social, 89
   as a "super-agent", 29

S

Second degree evolution, 119, 120
Self-awareness, 37

Self learning machines
  AlphaGo AI machines, 23–24
  AlphaGo Zero, 24–25
  deep neural network, 24
  development, 23
  "frontal teaching", 24
  interactions, 25, 26
  self-study, 24
  weak artificial intelligence, 25
Shadow economy
  AI machine, 67
  centauric intelligence, 70
  communication medium, 70
  digital companions, 68
  digital shadows, 67–70
  digital skin, 67
  digital twin, 67
  form of communication, 68
  human communication, 69
  hybrid intelligence, 68
  smartphone, 67
  worldwide learning community, 69
Shadow world, 68, 69
Smartphones, 9
  AI systems, 97
  communication, 98
  customers, 96
  digital companions, 10
  digital shadow, 68
  digital toolbox, 83
  embedded intelligence, 97
  logistics hub, 97
  "on demand", 96
  real shoe, 97, 98
  services, 98
  virtual shoe, 97, 98
Social bots
  computer program, 54
  "Deep Fake App", 55
  lip movements, 54
  Twitter world, 54
  two-dimensional space, 54

Social Core Process, 116
Social media, 49, 51, 54
Social robots, 89
Strong artificial intelligence
  A-consciousness, 37
  AI machine consciousness, 35
  "Big Data", 36
  consciousness, 35, 36
  data lakes, 36
  drones, 1
  legal entities, 12
  P-consciousness, 37
  self-awareness, 37
  systems, 35
Super Mario AI machine, 28

T

Teach-in learning, 24, 25, 32, 38–39
Thought consciousness (A-consciousness), 37
3D glasses, 10
3D printing, 96
Traditional business, 81–82
Transparent
  biological structure, 91
  logistics, 95
  robots, 30
  society, 56–57
  state of product, 91
Twitter, 53, 54, 56

U

Unobtrusive, 11, 70, 87

V

Value of agility, 113–115
Value of mindfulness, 115–116
Value of trust, 112–113
Viable System Model, 114
Voice machines, 72–74

W

Weak AI, 25, 27, 28, 88
WhatsApp, 50, 51, 53, 72
"Whereabouts", 48
Wireless communication, 75

Work-life balance, 48

Y

YouTube, 15, 49

GPSR Compliance

The European Union's (EU) General Product Safety Regulation (GPSR) is a set of rules that requires consumer products to be safe and our obligations to ensure this.

If you have any concerns about our products, you can contact us on

ProductSafety@springernature.com

In case Publisher is established outside the EU, the EU authorized representative is:

Springer Nature Customer Service Center GmbH
Europaplatz 3
69115 Heidelberg, Germany

www.ingramcontent.com/pod-product-compliance
Lightning Source LLC
LaVergne TN
LVHW010343260326
834688LV00036B/849